THE COMMONWEALTH AND INTERNATIONAL LIBRAF
OF SCIENCE, TECHNOLOGY, ENGINEERING
AND LIBERAL STUDIES

HYSIC

VOLUME ONE

General Editor
PROFESSOR J. T. WILSON

The Earth's Core and Geomagnetism

The Earth's Core and Geomagnetism

BY

J. A. JACOBS
M.A., Ph.D. D.Sc., F.R.A.S., F.R.S.C.

Director, Institute of Earth Sciences
University of British Columbia, Vancouver

PERGAMON PRESS
OXFORD · LONDON · PARIS FRANKFURT

THE MACMILLAN COMPANY
NEW YORK

PERGAMON PRESS LTD.
Headington Hill Hall, Oxford
4 & 5 Fitzroy Square, London W.1

THE MACMILLAN COMPANY
60 Fifth Avenue, New York 11, New York

COLLIER-MACMILLAN CANADA, LTD.
132 Water Street South, Galt, Ontario, Canada

GAUTHIER-VILLARS ED.
55 Quai des Grands-Augustins, Paris 6

PERGAMON PRESS G.m.b.H.
Kaiserstrasse 75, Frankfurt am Main

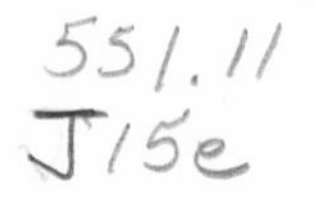

Library of Congress Card No. 63-21134

Set in 10 on 12 pt Times and printed in Great Britain by
PAGE BROS. (NORWICH) LTD.

Contents

CHAPTER 1

Introduction

DURING the past decade and particularly since the International Geophysical Year 1957–1958, there has been a tremendous increase in our knowledge of the Earth, its interior, the forces which shape its surface features, and its environment in space. Coupled with this increase of knowledge has come also an awareness of our own comparative ignorance of the way in which nature works. Opinions and theories of the Earth's interior are in a state of flux, which in itself is a healthy sign and indicative of the interest and amount of research which is going on in geophysics. Many statements about the Earth's interior which ten years ago were accepted as "practically certain" have since been found to be incorrect. Such fundamental problems as the origin of the Earth, its temperature distribution, its composition and continental drift, to name but a few, are just as controversial today as they were when they were first discussed.

No direct observations of the Earth's interior can be made below a depth of a few kilometres. Even when the Earth's mantle has been reached by drilling beneath the oceans where the Earth's crust is thinnest, less than one-fifth of 1 per cent of the distance towards the Earth's centre will have been reached. On the other hand laboratory experiments can only reproduce for any length of time the temperatures and pressures that exist in the outer few tens of kilometres of the Earth. Also it is almost impossible in any experimental work to make allowance for the enormous intervals of time involved in any geologic process. Thus if model work is carried out in an attempt to reproduce some of the effects observed in nature, the reduction in size imposed by a reasonable time scale usually makes the model microscopic. Conversely, the time scale imposed by a convenient choice of linear dimensions is in general far too great. However, electrical processes and mechanical motions do not scale down in the same way. An electromagnetic model of the Earth's core is difficult to make in the laboratory since if it is of reasonable size, the free decay time of any induced electric currents is far too *small* (see Section 6.1). Moreover in many cases restrictions imposed by correct scaling ratios require the model to be made of materials with unrealizable physical properties.

The fundamental problem in the physics of the Earth's interior is the determination of the correct rheological conditions which exist within the Earth. Rheology is the study of the deformation and flow of matter and different materials behave very differently under given conditions. The rheological behaviour of a material is determined by an equation of state involving stress, strain and possibly their time derivatives. Many rheological equations have been suggested all of which describe ideal bodies. Some of them, those for example which describe a perfectly elastic body or an ideal fluid, have been very highly developed mathematically but cannot explain all the characteristics of actual bodies especially when the stresses are applied for any length of time. More complex rheological equations to describe more sophisticated ideal bodies could be set up with but little advantage. The main difficulty is to determine the proper rheological conditions that exist in different parts of the Earth, and these may well be different for stresses applied over different time intervals. Such terms as "rigid" and "fluid" only have a meaning when the time interval over which the stresses are applied is specified.

Since it is not easy to reproduce in the laboratory conditions that prevail deep within the Earth, one can but hypothesize on the behaviour of materials below the crustal layers. The dangers of such hypotheses and perhaps self delusions based on an over-simplification of conditions that exist at depth within the Earth have been stressed by Professor F. Birch (1952), who has contributed so much to our knowledge of the Earth's interior.

"Unwary readers should take warning that ordinary language undergoes modification to a high-pressure form when applied to the interior of the Earth: a few examples of equivalents follow:

High-pressure form	*Ordinary meaning*
certain	dubious
undoubtedly	perhaps
positive proof	vague suggestion
unanswerable argument	trivial objection
pure iron	uncertain mixture of all the elements."

One of the major handicaps of a study of the deep interior of the Earth is the lack of experimental data so that the happy marriage of theory and experiment so fruitful in most branches of physics is often denied the geophysicist. Most information on the properties of the materials below the crustal layers is obtained by a theoretical interpretation of phenomena observed at the Earth's surface, and at great depths extrapolation of theoretical equations is frequently needed far beyond the conditions under which they were originally developed. Moreover the numerical values of physical constants are sometimes used far beyond the range of conditions under which they were obtained in the laboratory. An appreciation of many geophysical problems often requires a synthesis of knowledge from many disciplines such as astronomy, chemistry, geology, mathematics and physics. It is impossible for any one man to keep abreast of recent developments in any one of these fields, let alone in all of them. Thus a geophysicist may not be familiar with the underlying assumptions and limitations of a certain result he is using with the result that his extrapolation or extension of the theory may be grossly in error. To quote Professor B. Gutenberg (1959), "Conclusions concerning the deep portion of the mantle and the core may be subject to two major sources of error: those resulting from misinterpretation of observations, and those from application of theoretical equations which fit the problem only poorly or not at all and, in addition, may contain incorrectly estimated numerical factors." Moreover, scientists often have to make use of results in a field with which they are not familiar. Frequently they quote as well-established facts the results for "models" which had only been proposed as a working hypothesis. To quote Professor Gutenberg again, "It is nothing unusual that a tentative suggestion or model of a geophysicist is quoted by a writer in a different field as proof for one of his hypotheses"—creating the impression that such a tentative suggestion is in fact a definite result.

The mathematical development of a theory is frequently carried out to a much greater accuracy than is warranted by the imperfectly known or questionable assumptions of the physics of the process. Proper statistical analysis is essential in an interpretation of the results of many geophysical measurements, as for example in palaeomagnetism. However, if "probable errors" are calculated (by a least squares analysis or otherwise), then it does not necessarily follow that, if small, the result indicates confirmation of a theory—it may merely indicate good agreement between observations or the reproducibility of some experimental technique. Systematic misinterpretation of the results or incorrect assumptions may produce actual errors far greater than any calculated "probable errors".

The foregoing remarks are not meant to discourage the reader or to imply that we know nothing of the Earth's deep interior. It is, however, too easy to forget the limitations of our "knowledge". In studying geophysical phenomena we can in general observe conditions only at the moment, i.e. we can only record present values of quantities that have been varying for more than four billion years.* The development of isotope geophysics, the radioactive dating of rocks and minerals, and the determination of the remanent magnetism of rocks have made it possible to read some of the pages of the history of the Earth, but these more recent techniques are of little help in a study of the physics of the deep interior.

* 1 billion = 10^9 years. The age of the Earth is estimated to be about $4{\cdot}5 \times 10^9$ years.

References

BIRCH, F. Elasticity and constitution of the Earth's interior. *J. Geophys. Res.* **57**, 227–286 (1952).

GUTENBERG, B. *Physics of the Earth's Interior.* Academic Press (1959).

Suggestions for Further Reading

BASCOM, W. *A Hole in the Bottom of the Sea.* Doubleday, New York (1961).

HOWELL, B. F., JR. *Introduction to Geophysics.* McGraw-Hill (1959).

JACOBS, J. A., RUSSELL, R. D. and WILSON, J. T. *Physics and Geology.* McGraw-Hill (1959).

CHAPTER 2

Seismology and the Physics of the Earth's Interior

2.1 Elastic Waves

By far the most important source of information on the physical properties of the Earth's interior comes from seismology, i.e. the study of earthquakes, nature carrying out "experiments" in her own laboratory, experiments that man can but very imperfectly conduct himself. The Earth is not a static body but is continually undergoing deformation in response to the varying stresses that are set up within it. If the stresses in any region cannot be relieved (by elastic deformation or plastic flow), fracture may eventually take place. The sudden release of stress resulting from an earthquake will set up elastic waves issuing from a confined region below the Earth's surface, called the focus. The time during which the main energy is released is of the order of a second, and the linear dimensions of the focal region may be of the order of several kilometres. In large earthquakes seismic waves spread through the entire interior of the Earth, emerging again at the surface where they are recorded on seismograms at observatories all over the world. Elastic waves are, however, a more complex phenomenon than sound waves or waves propagated on the surface of the oceans.

Two distinct types of elastic waves can be propagated through the Earth, primary or P waves and secondary or S waves.* P

* Surface waves (i.e. waves travelling near the surface of the Earth, their amplitude decreasing rapidly with distance below the surface) will also be set up, and can give valuable information about the crustal layers. Long period surface waves can penetrate deep into the Earth's mantle and are beginning to yield information about the interior of the Earth.

waves, like sound waves, are longitudinal waves, and as the wave advances each particle of the solid is displaced in the direction of travel. S waves on the other hand are transverse waves, as are light waves, i.e. the motion of any particle is at right angles to the direction of travel of the waves. The analogy of P and S waves to sound waves and light waves is descriptive only of their mode of propagation and is not true of their velocity. The velocity of P waves is given by

$$V_P = \sqrt{\left(\frac{k + (4/3)\mu}{\rho}\right)} \tag{2.1}$$

and is in fact greater than the velocity of S waves which is given by

$$V_S = \sqrt{(\mu/\rho)} \tag{2.2}$$

In these formulae ρ is the density of the medium, k the bulk modulus or incompressibility, and μ the modulus of rigidity. It follows from equations (2.1) and (2.2) that

$$V_P^2 - \frac{4}{3}V_S^2 = \frac{k}{\rho} \tag{2.3}$$

Seismic evidence indicates that the Earth can be divided into three main divisions, the crust, the mantle and the core. The topmost layers of the Earth, called the crust, are of very variable thickness and composition. The continental crust consists of 30–60 km of light rocks (such as gneiss, granodiorite and granite), while the oceanic crust is usually no more than 5–6 km thick and is made up of darker rocks (such as basalt). The mantle lies between the crust and a depth of approximately 2900 km and the region below the mantle is called the core. There are sub-divisions of these three main divisions, the structure of the crust in particular being extremely complex in some parts of the world. More details of the deeper interior of the Earth will be given later in this chapter. In particular there is evidence that the core may be divided into two parts, an inner core and an outer core. S waves have never been observed in the core, and this is the main reason for believing the core to be liquid, since a liquid cannot transmit shear waves (if $V_S = 0$, equation (2.2) shows that $\mu = 0$). There is some evidence, however, that the inner core may be solid and thus able to transmit S waves. This controversial question will be discussed again in Chapters 3 and 4.

A seismogram may show many wave phases, each phase corresponding to a particular route of a ray from the focus to the seismic station, many routes being possible because of reflection at the Earth's surface and reflection and refraction in the interior. When an elastic wave strikes a discontinuity such as the surface of the Earth or the boundary between the core and the mantle, it will be both reflected and refracted, and laws of reflection and refraction analogous to those of geometrical optics apply. But here the analogy ends, for both P and S waves may be reflected and refracted when an incident wave (of either P or S type) strikes a boundary. Thus an incident P wave reflected from the surface of the Earth gives rise to both a P wave, called PP, and an S wave, called PS. Similarly an incident S wave can be reflected to give both SP and SS waves. Waves which strike the boundary between the mantle and the core may be both reflected and refracted. The symbol c is used to indicate an upward reflection at the core boundary. Thus ScP is a shear wave that has travelled down to the core boundary and been reflected as a P wave. As already mentioned no shear waves have ever been observed in the core, although there is some doubt about the rigidity of the inner core. The symbol K is used to denote a wave (necessarily of P type) propagated through the outer core. Thus SKS is a shear wave that has been refracted into the outer core (as a P wave) and refracted back into the mantle as an S wave again. Because of our uncertainty about the rigidity of the inner core, the symbol I is used to describe a P wave in the inner core and the symbol J for S waves (if they exist). Thus PKIKP is a P wave which has passed from the mantle through both the outer and inner core and back into the mantle again, remaining a P wave throughout its path. Figure 2.1, after Professor K. E. Bullen, one of the pioneers in this field, illustrates the complexity arising from the great number of reflections and refractions of elastic waves that may result from an earthquake.

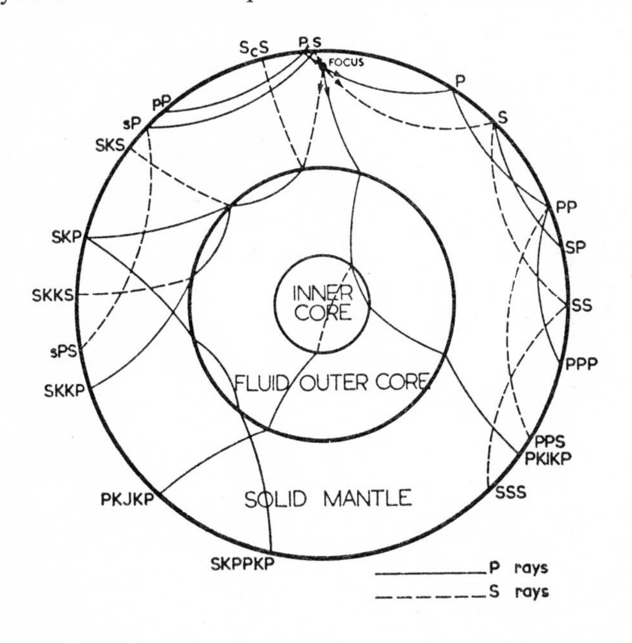

FIG. 2.1 Representative seismic rays through the Earth (after K. E. Bullen).

2.2 Travel–time Curves and Velocity–depth Curves

The times of arrival of elastic waves generated by an earthquake can be determined on the seismograms at a number of stations, so that it is possible to construct travel–time curves. These are plots of arrival times against the distance Δ (measured in degrees along the surface of the Earth), between the source and the seismic observatory. Figure 2.2 gives an example of travel–time curves constructed by Sir Harold Jeffreys and Professor K. E. Bullen for a number of different wave types. From such travel–time

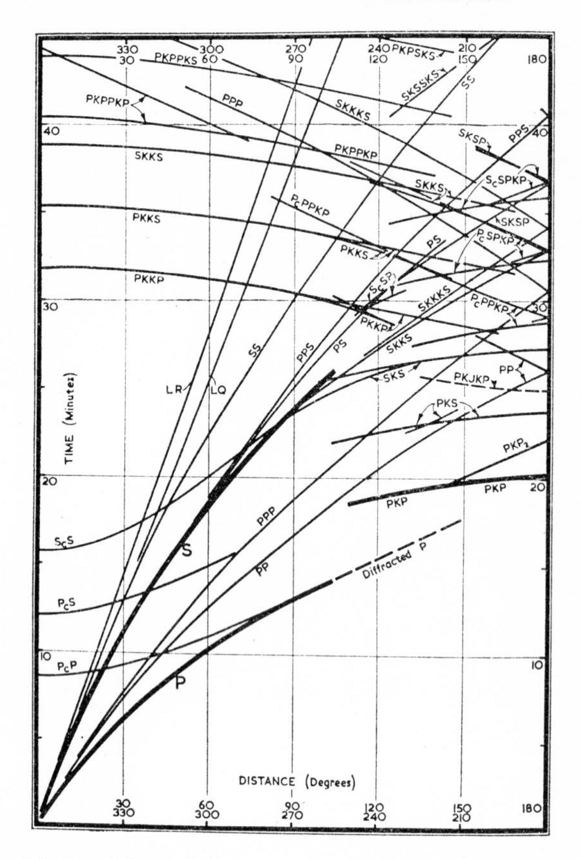

FIG. 2.2 The Jeffreys–Bullen travel–time curves (after K. E. Bullen).

curves it is possible to calculate the velocity at any depth within the Earth. The calculations are, however, not easy and will not be given here. Estimates of velocity–depth curves have been made since 1910 onwards and Fig. 2.3 gives two of the most recent determinations by Sir Harold Jeffreys and Professor B. Gutenberg. Slight differences in these two estimates exist in the upper part of the mantle but they do not affect the results deeper in the Earth and thus will not be discussed in detail here. On the other hand, there is a difference in interpretation at the boundary of the inner core at a depth of approximately 5120 km. Jeffreys postulates a major discontinuity, whereas Gutenberg concludes that there is only a sharp change of slope in the velocity–depth curve and does not believe that there is a region of decreasing velocity with

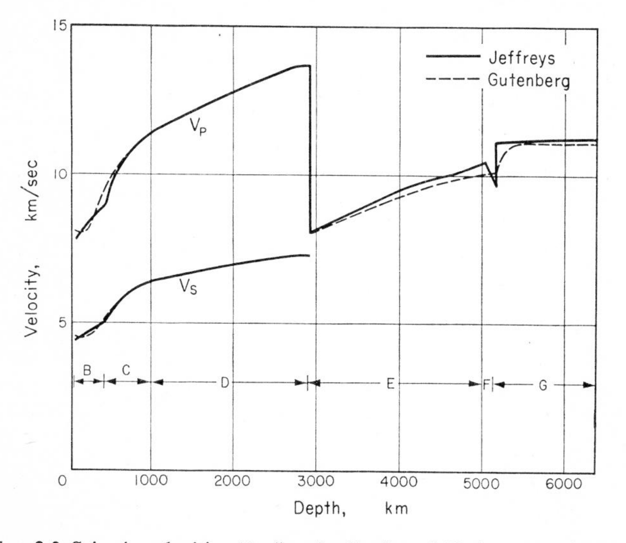

FIG. 2.3 Seismic velocities V_P (longitudinal) and V_S (transverse) as a function of depth (after F. Birch).

depth nor a discontinuous increase. Table 2.1 gives the velocities of P and S waves at different depths within the Earth according to Jeffreys. The most recent tables of transmission times t of longitudinal waves through the Earth as a function of Δ usually agree within 3 sec in values up to over 1000 sec. However, when these raw data are used in calculations the magnitude of possible errors may increase with the number of mathematical operations.

The velocity–depth curves of Fig. 2.3 provide the initial data from which much of our knowledge of the interior of the Earth

TABLE 2.1. *Velocities of P and S Waves at Different Depths in the Earth (According to H. Jeffreys)*

Depth km	Velocity V_P of longitudinal (P) waves, km/sec	Velocity V_S of transverse (S) waves, km/sec	Depth km	Velocity V_P of longitudinal (P) waves, km/sec
33	7·75	4·35	2898	8·10
100	7·95	4·45	3000	8·22
200	8·26	4·60	3200	8·47
300	8·58	4·76	3400	8·76
413	8·97	4·96	3600	9·04
600	10·25	5·66	3800	9·28
800	11·00	6·13	4000	9·51
1000	11·42	6·36	4200	9·70
1200	11·71	6·50	4400	9·88
1400	11·99	6·62	4600	10·06
1600	12·26	6·73	4800	10·25
1800	12·53	6·83	4892	10·44
2000	12·79	6·93	5121	(9·7)
2200	13·03	7·02	5121	11·16
2400	13·27	7·12	5700	11·26
2600	13·50	7·21	6371	11·31
2800	13·64	7·30		
2898	13·64	7·30		

is based. From an analysis of these curves, Bullen has designated a number of sub-divisions of the Earth which are given in Table 2.2. The most striking feature of the velocity–depth curves is the discontinuity at a depth of 2898 km which marks the boundary between the mantle and the core. There is a discontinuous drop in the velocity of P waves, and S waves have never been detected below that depth, which is the main reason for believing that the core is liquid. H. Takeuchi using tidal data and seismic data in the mantle has shown that the rigidity of the outer core E is less than one-fortieth of the rigidity of the rocks of the outer mantle, again indicating that the region E is essentially fluid, i.e. a material of zero or very small rigidity. The physical state of the inner core is, however, still in doubt. It is with the core of the Earth where

TABLE 2.2. *Dimensions and Descriptions of Internal Layers of the Earth (After K. E. Bullen)*

Layer		Depth to boundaries, km	Fraction of volume	Features of regions
Crust*	A	0	0·0155	Conditions fairly heterogeneous
		33		
	B		0·1667	Probably homogeneous
		413		
Mantle	C		0·2131	Transition region
		984		
	D		0·4428	Probably homogeneous
		2898		
	E		0·1516	Homogeneous fluid
		4982		
Core	F		0·0028	Transition layer
		5121		
	G		0·0076	Inner core (solid?)
		6371		

* The thickness of the crust under the continents is not constant; it averages about 33 km. Under the oceans, the crust is much thinner, being only about 5 or 6 km thick.

the Earth's magnetic field is believed to originate, that this book is mainly concerned. There is a second major discontinuity in the velocity–depth curves just below the Earth's surface which marks the boundary between the crust and the mantle. This boundary is called the Mohorovičič discontinuity after its discoverer who identified it on a seismogram of an earthquake in Croatia in 1909. The depth to the "Moho" is about 35 km under the continents and about 5 km under the ocean floors. Although the detailed structure of the crust is extremely complex, it is less than 1 per cent of the mass of the whole Earth and the details of its structure do not affect the physical properties of the deeper parts of the mantle and core. In calculations involving the deep interior of the Earth the crust has been conventionally taken to consist of a layer 15 km thick and of density 2·65 g/cm^3 resting on a second layer 18 km thick and of density 2·87 g/cm^3. This model is now out of date but, since there is no definite crustal picture as yet, it is advantageous to keep this model for standard comparison purposes in calculations relating to the whole Earth. Errors in a model of the crustal layers have little effect on results in the deep interior. There are three other possible discontinuities in the Earth, none of them so pronounced as the core–mantle boundary or the Moho. One is the boundary between the inner and outer cores at a depth of 5121 km. The other two are discontinuities in the slope of the velocity–depth curves in the upper mantle and have been placed by Bullen at depths of 413 and 984 km. It must be stressed again that the velocity–depth curves in Fig. 2.3 are the interpretation of the seismic data by two of the leading experts in this field, yet nevertheless they are only models. Although it is unlikely that there will be any major revisions, the detailed structure of the curves is not well-established—for example there is considerable doubt that there is a real discontinuity at 413 km, and possible discontinuities have been suggested near depths of 200 km and 500 km.

Evidence for the existence of an inner core is given by the occurrence of a "shadow zone" for P waves emerging at distances Δ between 105° and 142°. A PKP wave emerging at $\Delta = 142°$ has the least value of Δ for a wave which penetrates the outer but not the inner core (see Fig. 2.4). The range $105° < \Delta < 142°$ is not a true shadow zone since some waves of P type are observed in it. The amplitudes of such waves are much reduced and for many years their presence was attributed to diffraction round the boundary of the core. This explanation is satisfactory for waves recorded at distances a little greater than 105°, but cannot account for waves received at the other end of the shadow zone. Miss I. Lehmann suggested in 1935 that the waves recorded in the shadow zone had passed through an inner core in which the P velocity is significantly greater than that in the outer core. Later work has corroborated her hypothesis and the existence of an inner core seems fairly well established now. The physical state of the inner core and whether it is solid or liquid will be discussed in Chapters 3 and 4.

B. A. Bolt (1962) has revised the Jeffreys distribution of the velocity of P waves in the Earth's core, obtaining two discontinuous increases in V_P at the boundaries of an intermediate region F separating an inner core G from an outer core E. The velocities throughout F and G are taken as constant. Bolt made this revision in an attempt to explain the first arrivals PKP at epicentral distances Δ between about 125° and 140°. At the shorter end of this range such waves arrive some 15–20 sec

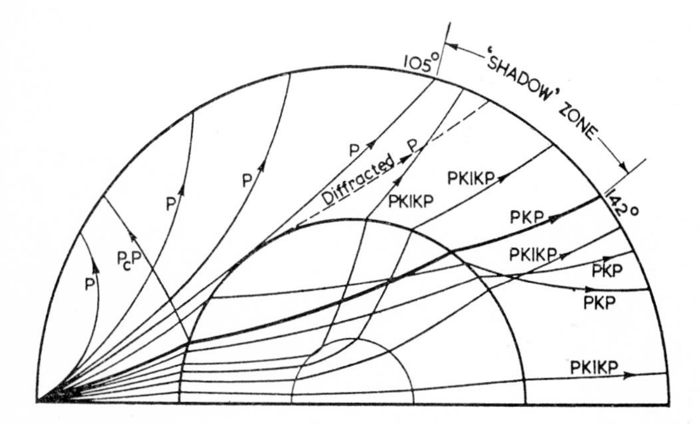

FIG. 2.4. P, PcP, PKP, PKIKP and diffracted P waves (after K. E. Bullen).

earlier than the normal larger amplitude PKIKP waves. Gutenberg had already called attention to these early onsets of PKP waves and put forward as a tentative explanation wave dispersion in the transition region F. Bolt's interpretation with its revised velocity–depth curve has an important bearing on the physical properties of the Earth's core (see Section 2.3).

2.3 Variation of Density and other Physical Properties within the Earth

The variation of the density ρ of the Earth cannot be found uniquely as a function of the radius r from observations at the surface— all that we can do is to construct possible Earth models. However, in addition to the seismic data which provide values of k/ρ and μ/ρ (see equations 2.3 and 2.2), there are other conditions that any function $\rho(r)$ must satisfy, and Bullen has shown that in much of the Earth, the permissible variations of $\rho(r)$ are quite small.

It is usually assumed that the variation of pressure p is given by the hydrostatic equation

$$\frac{\mathrm{d}p}{\mathrm{d}r} = -g\rho \tag{2.4}$$

where

$$g = Gm/r^2 \tag{2.5}$$

G is the constant of gravitation and m the mass of material within a sphere of radius r. The assumption of hydrostatic pressure does not hold in the crust, but the mean pressure increases with depth while the maximum stress difference decreases and a depth is soon reached when equation (2.4) is sufficiently accurate. In a homogeneous layer where the temperature variation is adiabatic, it follows from equation (2.4) that

$$\frac{\mathrm{d}\rho}{\mathrm{d}r} = \frac{\mathrm{d}\rho}{\mathrm{d}p} \cdot \frac{\mathrm{d}p}{\mathrm{d}r} = \frac{\rho}{k} \cdot (-g\rho) = \frac{-g\rho^2}{k} \tag{2.6}$$

where k is the (adiabatic) incompressibility. By a homogeneous region is meant one in which there are no significant changes of phase or chemical composition. If the temperature variation is not adiabatic, the right-hand side of equation (2.6) may be modified by including a factor $(1 - \delta)$, i.e.

$$\frac{\mathrm{d}\rho}{\mathrm{d}r} = \frac{-g\rho^2}{k}(1 - \delta) \tag{2.7}$$

F. Birch (1952) has estimated that a variation of 1 °C/km from the adiabatic gradient may affect the value of the density gradient by as much as 10 per cent. There is no direct evidence of the variation of the temperature gradient in the Earth's deep interior, although more recent calculations tend to reduce earlier estimates of differences from the adiabatic gradient. Since the effect of the term δ in equation (2.7) is to increase $d\rho/dr$ whereas the effect of any variance from chemical homogeneity is in the opposite direction, δ is taken as zero, i.e. equation (2.6) has been applied in those regions where there is no evidence of inhomogeneity. The detailed analysis of Birch (1952) is compatible with neglecting δ at depths greater than about 1000 km although the term may be important in the upper mantle. Bullen estimates that neglecting δ should not lead to errors in the computed densities exceeding about 0·1 g/cm³.

Combining equations (2.5) and (2.6), we have

$$\frac{d\rho}{dr} = \frac{-Gm\rho^2}{kr^2} \tag{2.8}$$

Since $dm/dr = 4\pi\rho r^2$ and k/ρ is known from the velocity–depth curves (see equation 2.3), equation (2.8) can be integrated numerically to obtain the density distribution in regions of the Earth where chemical and non-adiabatic temperature variations may be neglected. This equation was first obtained by L. H. Adams and E. D. Williamson in 1923 and has been extensively used by Bullen and later by E. C. Bullard, W. H. Ramsey and B. A. Bolt to obtain density distributions.

Any density distribution must satisfy two conditions—it must yield the correct total mass M of the Earth ($5{\cdot}977 \times 10^{27}$ g) and the correct moment of inertia I about its rotational axis ($0{\cdot}334Ma^2$, where a is the mean radius). Using these two conditions and a value ρ_1 of 3·32 g/cm³ for ρ at the top of layer B of the mantle, Bullen applied equation (2.8) throughout layers B, C, and D. He found that this solution lead to a value of the moment of inertia I_c of the core greater than that of a uniform sphere of the same size and mass. This would entail the density decreasing inwards which would be an unstable state in a fluid. Allowance for the term δ (i.e. the use of equation 2.7 instead of 2.8) increased the value of I_c. A reasonable value could be obtained by increasing the value of ρ_1—but only if an impossibly high value is taken. Hence the assumption of chemical homogeneity must be in error, i.e. the Adams–Williamson equation (2.8) cannot be used throughout the mantle. The most likely region where this assumption breaks down is C where the velocity–depth curves indicate changes in slope. In his Earth Model A Bullen thus used equation (2.8) in regions B and D while in C he fitted a quadratic expression in r for $\rho = \rho(r)$. In the outer core (region E) equation (2.8) is likely to apply, and values of ρ down to a depth of about 5000 km can be obtained with some confidence, although it is not easy to determine the density in regions F and G with any certainty. However since these regions constitute only about 1 per cent of the Earth's total volume, the density distribution within E can be estimated fairly precisely. Sir Edward Bullard (1957) has shown that an increase of 1 g/cm^3 in the density at the surface of the inner core G would reduce the density in the outer core by only 0·07 g/cm^3. Figure 2.5 shows the density distribution obtained by Bullen for two Earth models A and B. The details of these and other models in so far as they affect the core will be discussed later.

Bullen (1962) has derived an expression η for a measure of the departure from chemical homogeneity at any depth z in the Earth, viz.

$$\eta = \frac{\mathrm{d}k}{\mathrm{d}p} - \frac{1}{g}\frac{\mathrm{d}}{\mathrm{d}z}\left(\frac{k}{\rho}\right) \tag{2.9}$$

η is the ratio of the actual value of $\mathrm{d}\rho/\mathrm{d}z$ to the value it would take if the composition were unvarying. η is equal to unity if the Earth is chemically homogeneous, but is greater than one where the chemical composition is varying with depth. Using Jeffrey's velocity–depth curves, η is of the order of 30 in the region F

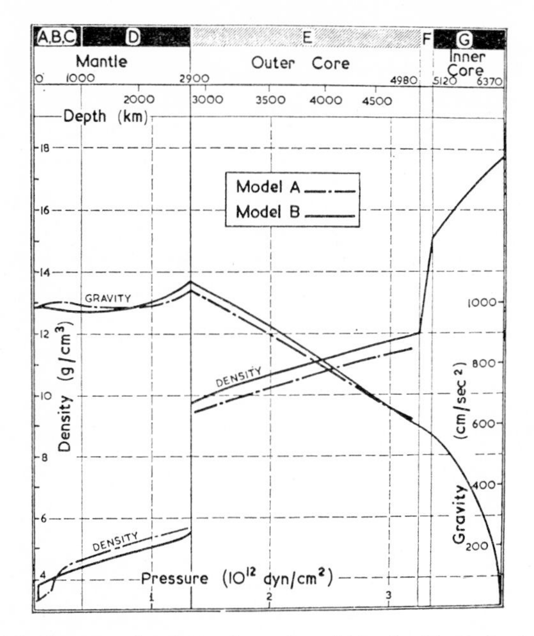

FIG. 2.5. Pressure, density and gravity within the Earth for Models A and B (after K. E. Bullen).

implying a large density increase (of the order of 3 g/cm³) resulting in an estimated density of the order of 18 g/cm³ at the centre of the Earth. Using Bolt's (1962) revised distribution of the velocity of P waves in the core, the second term in equation (2.9) is zero in F (since V_P is constant throughout F), and the value of η is reduced from 30 to about 4 or 5. This implies a substantial reduction in the estimate of the Earth's central density—of the order of 2 g/cm³ at least, and a figure of 16 g/cm³ (or possibly a little less) is now preferred for the central density.

From equations (2.4) and (2.5), it follows that

$$\frac{dp}{dr} = \frac{-Gm\rho}{r^2} \tag{2.10}$$

so that the pressure distribution may be obtained by numerical integration once the density distribution has been obtained. The pressure distribution is insensitive to small changes in the density distribution, since the density is used only to determine the pressure gradient. The variation of g can be obtained from equation (2.5). The distributions of pressure and g for Bullen's Earth models A and B are also given in Fig. 2.5, and Tables 2.3 and 2.4 give the numerical values of the physical properties of model A.

TABLE 2.3. *Density, Gravity, and Pressure Distribution in the Mantle: Earth Model A (After K. E. Bullen)*

Depth, km	Density, g/cm^3	Gravity, cm/sec^2	Pressure, $\times 10^{12}$ dynes/cm^2
33	3·32	985	0·009
100	3·38	989	0·031
200	3·47	992	0·065
300	3·55	995	0·100
400	3·63	997	0·136
413	3·64	998	0·141
500	3·89	1000	0·173
600	4·13	1001	0·213
700	4·33	1000	0·256
800	4·49	999	0·300
900	4·60	997	0·346
1000	4·68	995	0·392
1200	4·80	991	0·49
1400	4·91	988	0·58
1600	5·03	986	0·68
1800	5·13	985	0·78
2000	5·24	986	0·88
2200	5·34	990	0·99
2400	5·44	998	1·09
2600	5·54	1009	1·20
2800	5·63	1026	1·32
2898	5·68	1037	1·37

Finally from a knowledge of the density distribution it is possible to compute the variation of the elastic constants.

TABLE 2.4. *Density, Gravity, and Pressure Distribution in the Core: Earth Model A (After K. E. Bullen)*

Depth, km	Density, g/cm^3	Gravity, cm/sec^2	Pressure $\times 10^{12}$ $dynes/cm^2$
2898	9·43	1037	1·37
3000	9·57	1019	1·47
3200	9·85	979	1·67
3400	10·11	936	1·85
3600	10·35	892	2·04
3800	10·56	848	2·22
4000	10·76	803	2·40
4200	10·94	758	2·57
4400	11·11	716	2·73
4600	11·27	677	2·88
4800	11·41	646	3·03
4982	11·54	626	3·17
5121	(14·2)	585	3·27
5121	(16·8)	—	—
5400	—	460	3·41
5700	—	320	3·53
6000	—	177	3·60
6371	(17·2)	0	3·64

Equations (2.2) and (2.3) give μ and k directly. From the known relations between the elastic constants, it is possible to compute the distribution of Young's modulus E and Poisson's ratio σ. In particular

$$\sigma = \frac{3k - 2\mu}{6k + 2\mu}$$

and is thus independent of the density ρ. Figure 2.6 illustrates the variation of μ and k throughout the Earth for Bullen's two Earth models A and B—values in the inner core are rather uncertain.

F. Birch (1961), assuming a mean atomic weight throughout the mantle and a linear relationship between velocity and density for large compression, has obtained a density distribution very similar to Bullen's Model A. He believes that allowance for the effect of temperature is likely to reconcile all of the data with a mantle of nearly uniform mean atomic weight close to that of average chondritic material.

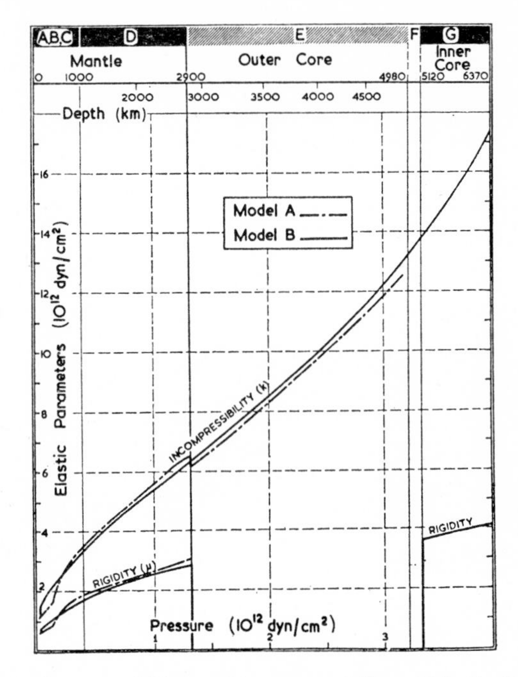

FIG. 2.6. Pressure, rigidity and incompressibility within the Earth for Models A and B (after K. E. Bullen).

E. C. Bullard (1957) has carried out on an electronic computer a large number of integrations of the Adams–Williamson equation (2.8) in order to determine the uncertainties in the density distribution of errors in the assumptions. He showed that the only sources of serious error are the uncertainties about the distribution of inhomogeneity and temperature in the mantle. The effects of these far exceed those due to uncertainties in the seismic velocity distributions and a possible discontinuity at the surface of the inner core. If the temperature distribution is adiabatic, Bullen's solution with the inhomogeneity spread between 413 and 984 km below the surface should not be in error by more than $\pm 0{\cdot}3$ g/cm^3 in the mantle and $\pm 0{\cdot}5$ g/cm^3 in the core. If the temperature is near the melting point, densities deep in the mantle may be reduced by 0·2 g/cm^3 and those in the core raised by about the same amount.

References

BIRCH, F. Elasticity and constitution of the Earth's interior. *J. Geophys. Res.* **57**, 227–86 (1952).

BIRCH, F. Composition of the Earth's mantle. *Geophys. J.* **4**, 295–311 (1961).

BOLT, B. A. Earth models with continuous density distribution. *Mon. Not. Roy. Astr. Soc. Geophys. Suppl.* **7**, 360–8 (1957).

BOLT, B. A. Gutenberg's early PKP observations. *Nature* **196**, 122–4 (1962).

BULLARD, E. C. The density within the Earth. *Verhandel Ned. Geol. Mijnbouwk-Genoot.* (*Geol. Ser.*) **18**, 23–41 (1957).

BULLEN, K. E. *Introduction to the Theory of Seismology.* Cambridge Univ. Press (1953).

BULLEN, K. E. *Seismology.* Methuen (1954).

BULLEN, K. E. Earth's central density. *Nature* **196**, 973 (1962).

LEHMANN, I. *Publ. Bur. Cent. Seism. Int. Ser. A* **14**, 87–115 (1936).

RAMSEY, W. H. On the constitution of the terrestrial planets. *Mon. Not. Roy. Astr. Soc.* **108**, 406–13 (1948).

RAMSEY, W. H. On the nature of the Earth's core. *Mon. Not. Roy. Astr. Soc. Geophys. Suppl.* **5**, 409–26 (1949).

TAKEUCHI, H. On the earth tide in the compressible Earth of varying density and elasticity. *Trans. Amer. Geophys. Union* **31**, 651–89 (1950).

WILLIAMSON, E. D. and ADAMS, L. H. Density distribution in the Earth. *Washington Acad. Sci. J.* **13**, 413–28 (1923).

CHAPTER 3

Constitution and Composition of the Core

3.1 Ramsey's Hypothesis

Estimates of the density distribution within the Earth have been given in Chapter 2. The mean density of the Earth (about 5·5 g/cm^3) is almost twice that of surface rocks (about 2·8 g/cm^3), so that even allowing for the compression of materials due to the increased pressure at depth within the Earth, the density distribution cannot be explained by a composition similar to that of surface rocks. Although the effects of high pressure are not well known, seismic data indicate that the Earth has a liquid core with a density several times that of silicate rocks. This situation has been compared to that observed in the smelting of iron in blast furnaces, where the iron, reduced to a metallic state, sinks to the bottom where it forms a single, dense, liquid phase, the residual silicates floating to the surface as slag. Four elements, iron, oxygen, magnesium and silicon make up more than 90 per cent of the bulk composition of the Earth. Part of the iron may have combined to form iron or iron–magnesium silicates such as olivine, the chief constituent of dunite and peridotite, the rest remaining in a metallic state. Free iron and the iron silicates form two immiscible phases, and the heavier iron phase would settle to the centre of the Earth under the action of gravity. This model of the Earth has been strengthened by the presence of silicate and metallic nickel–iron phases in meteorites. Professor H. C. Urey (1952) has suggested that the Earth's core has not always been of the same size but has grown throughout geologic time, iron continuing to sink through the silicate mantle. The moment of inertia of the Earth would thus have decreased with time, leading to an increase in its rate of rotation, i.e. a decrease in the length of the day.

This model of an iron or iron–nickel core surrounded by a silicate mantle has been challenged by W. H. Ramsey (1948) who suggested that the discontinuity at a depth of 2898 km is due to a high-pressure phase change. Ramsey thus considers the Earth to be chemically homogeneous (below the crustal layers), the core–mantle boundary representing a transition from a molecular to a metallic phase. Ramsey originally put forward this hypothesis to account for the densities of the inner planets which he assumed to have a common primitive composition. More recent astronomical data, however, shows that his hypothesis is unlikely to be true and that the outer core most probably contains some uncombined iron and also material of lower atomic number —perhaps in the form of a phase transformation of the material of the lower mantle. Professor P. W. Bridgman has accumulated a large amount of experimental data on the densities and compressibilities of materials up to 10^5 atm. On the other hand theoretical data at pressures of 10^7 atm and above may be obtained using a quantum statistical method, based on a Thomas–Fermi–Dirac model for the electrons surrounding the nucleus. In this model the electronic shells of the atoms are pressed together and lose their individual structure. Many investigators have attempted to obtain information relevant to conditions in the Earth's interior by interpolating in this pressure range. Their results will be discussed below, but they do not support Ramsey's hypothesis. Finally a phase change in a multicomponent system almost certainly will be spread out over a range of pressures, whereas the core–mantle boundary presents a very sharp discontinuity. Although Ramsey's suggestion is not generally accepted now, it engendered a considerable amount of controversy at the time and it has undoubtedly helped to increase our knowledge of the Earth's interior.

3.2 Bullen's Compressibility-pressure Hypothesis

From the results of his Earth Model A, K. E. Bullen found that there was no noticeable difference in the incompressibility gradient dk/dp between the base of the mantle and the top of the core. Moreover there was only a 5 per cent difference in the value of k across the core–mantle boundary. These features are in marked contrast to the large changes in the density and rigidity at that boundary. The change in k is a diminution from the mantle to the core. However, interpolation between experimental data at 10^5 atm and theoretical studies at 10^7 atm, indicates that in the transition from the mantle to the core, a slight increase in k could be expected for materials likely to occur in the Earth's interior. Because of the smallness in the change in k across the core–mantle boundary and because this change is opposite in sign to that predicted by such an interpolation, Bullen (1949, 1950) proposed another Earth Model B in which he assumed that k and dk/dp are smoothly varying functions throughout the Earth below a depth of about 1000 km. This hypothesis, called the compressibility pressure hypothesis, implies that at high pressures the compressibility of a substance is independent of its chemical composition. More recent work in theoretical physics indicates that the hypothesis as stated is a little too general and that there is some small variation of k with atomic number at high pressures. However these variations are likely to be small and the hypothesis is probably a good approximation. Bullen has shown that it is highly probable that the density and compressibility values in the actual Earth in the regions D and E lie between the Model A and Model B values. Sir Edward Bullard (1957) has investigated permissible density distributions within the Earth and has concluded that Bullen's estimate of 9·7 g/cm^3 at the core–mantle boundary should not be in error by more than 0·5 g/cm^3.

The velocity distribution of P waves according to Jeffreys has a negative gradient in an intermediate region F between the outer core E and the inner core G (see Fig. 2.3). The evidence for the existence of the F layer rests on two earthquakes together with the accumulated data for travel–time curves from many other earthquakes. It is to be noticed that Gutenberg's determination of the velocity–depth curves does not show the existence of an F layer. According to Jeffreys, the velocity V_P of P waves diminishes from a value of 10·44 km/sec at the top of the F layer to a value of 9·47 km/sec at the bottom. This is followed by a discontinuous jump from 9·47 to 11·16 km/sec across the boundary between F and G. If the layers E and F are both liquid so that $V_S = 0$, then from equation (2.3), V_P is given by

$$V_P^2 = k/\rho \tag{3.1}$$

and it follows that either dk/dp is negative throughout F or else ρ increases by at least 16 per cent through F. Moreover, if equation (3.1) holds in the inner core G (i.e. if the inner core is liquid), k would have to increase by 32 per cent across the boundary between F and G, excluding the highly improbable case that the density decreases with depth. On the other hand, if the inner core G is solid, and thus capable of transmitting P waves, equation (3.1) is replaced by equation (2.3) in G. Thus the discontinuity in the value of k across the boundary between F and G can be avoided if at the top of G,

$$\frac{4}{3}V_S^2 = (11{\cdot}16^2 - 9{\cdot}47^2)\ \text{km}^2/\text{sec}^2$$

which gives a value of 4·8 km/sec for V_S inside G.

On Gutenberg's interpretation of the velocity–depth curve in the core, the effective increase in V_P^2 between F and G is 23 per cent. Thus, since it is almost impossible that ρ should decrease with depth inside the core, both interpretations indicate that the inner core has significant rigidity comparable with that in the lower mantle. A sharp boundary such as would be provided by a solid inner core would also make it easier for convection currents to occur in the fluid outer core. Such currents will be invoked in Chapter 5 in discussing the Earth's magnetic field.

3.3 Experimental and Theoretical Work at High Pressures

It is always desirable to confirm by experiment any theoretical arguments, but this is often denied the geophysicist. Not only does the scale of natural phenomena, both in dimensions and time, make it almost impossible to carry out laboratory experiments, but it is also extremely difficult to reproduce the conditions of pressure and temperature that exist deep within the Earth. The pioneering experimental work of Professor P. W. Bridgman up to pressures of 10^5 atm corresponds to a depth of only 300 km within the Earth. However in the past few years, dynamic determinations of the compressibility of metals have been made (L. V. Altshuler *et al.*, 1958; R. G. McQueen and S. P. Marsh, 1960) up to pressures of 5×10^6 atm which is greater than the pressure at the centre of the Earth.

These high pressures are created for very short time intervals behind the front of a strong shock wave set up by an explosive charge, and are an order of magnitude greater than pressures which can be obtained by static methods.

In order to interpret the data, the equation of state as determined from the shock wave data, which is neither adiabatic nor isothermal, must be reduced to a reference temperature. Temperatures in the shock front are not generally known and additional measurements must be made to reduce the pressure–density data to those at absolute zero. The equation of state for iron (reduced to absolute zero) is reproduced in Fig. 3.1 and shows that the experimental data give a density greater than the density of the core at pressures existing at the core–mantle boundary. The portion of the curves in Fig. 3.1 appropriate to the core of the Earth are replotted on a different scale in Fig. 3.2. If no corrections are applied, the experimental data indicate that the density of iron would be of the order of 1·8 g/cm³ greater than that in the Earth's core. L. Knopoff and G. J. F. MacDonald (1960) have shown that this difference can be reduced but still exists even when corrections are made to the experimental data to allow for the thermal expansion of iron to the temperatures of the core and for

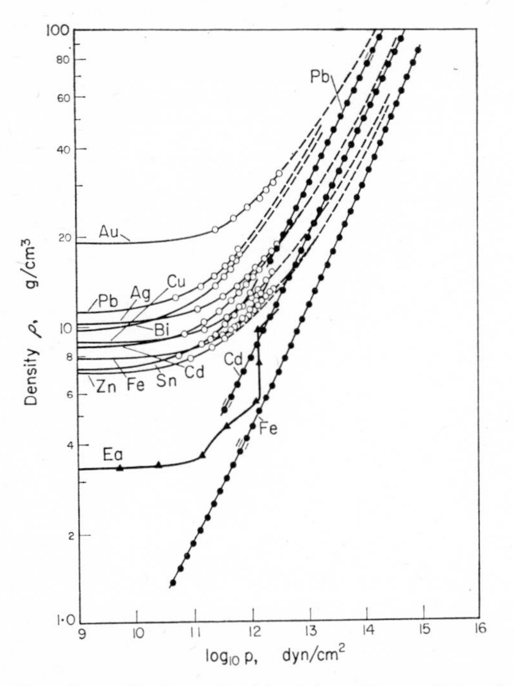

FIG. 3.1. Experimentally determined equation of state of nine elements (open circles) reduced to absolute zero, compared with the Thomas–Fermi equation of state at absolute zero (closed circles) and the equation of state for the Earth on Bullen's model (triangles). Extrapolation of the experimental data (dashes) is obtained by integration of the velocity equation of state (after L. Knopoff and G. J. F. MacDonald).

the volume change upon melting. The discrepancy can only be resolved if the core is not pure iron but contains significant amounts of elements of lower atomic number.

L. Knopoff and R. J. Uffen (1954) have extended the quantum-statistical method for computing the densities of pure elements at very high pressures to solid compounds. They have then interpolated in the interval between Bridgman's experimental data and these theoretical predictions to obtain pressure–density

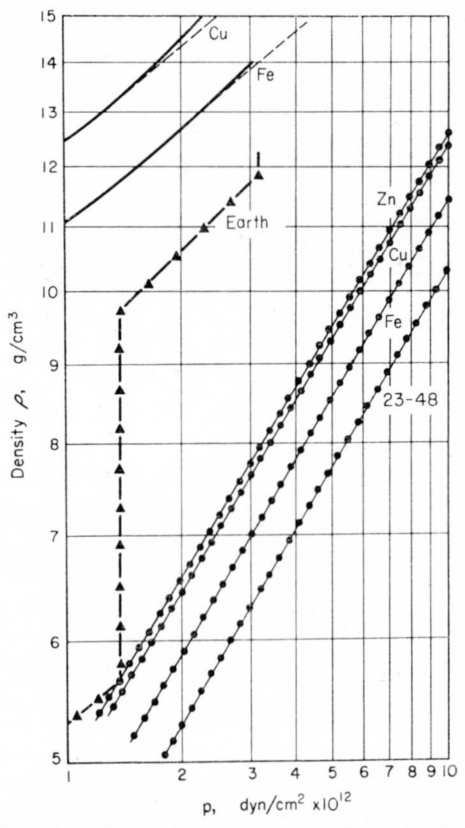

FIG. 3.2 Equation of state for iron, copper, zinc and a hypothetical material of atomic number 23, atomic weight 48 in the core pressure range. The values derived from shock-wave measurements (solid) are compared with those obtained from the Thomas–Fermi theory (closed circles). Bullen's density distribution is shown for comparison (triangles) (after L. Knopoff and G. J. F. MacDonald).

curves (see Fig. 3.3) for all probable constituents of the deep interior of the Earth. They improved the interpolation by extending the high pressure data of Bridgman using the Birch–Murnaghan semi-empirical theory of finite strain. The quantum

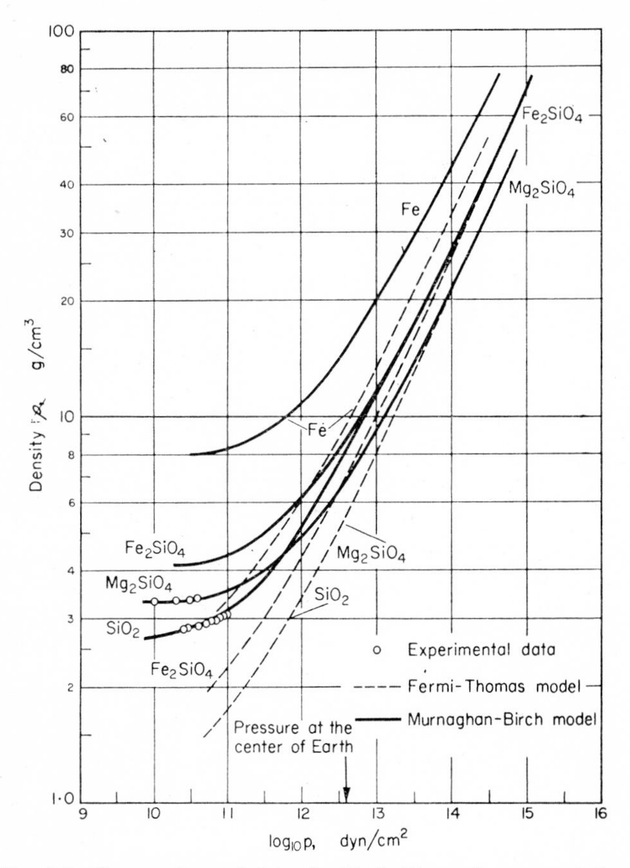

FIG. 3.3. The experimental data, the Birch–Murnaghan computation, and the Fermi–Thomas computations for four materials (after L. Knopoff and R. J. Uffen).

calculations begin to hold at pressures of the order of 10^{14} dynes/cm^2, pressures greatly exceeding those within the Earth. Figure 3.4 shows the pressure–density relation for the Earth (using Bullen's results) as compared with the interpolated curves for fayalite, forsterite, iron and nickel. The quantum method is strictly speaking only valid for temperatures at absolute zero.

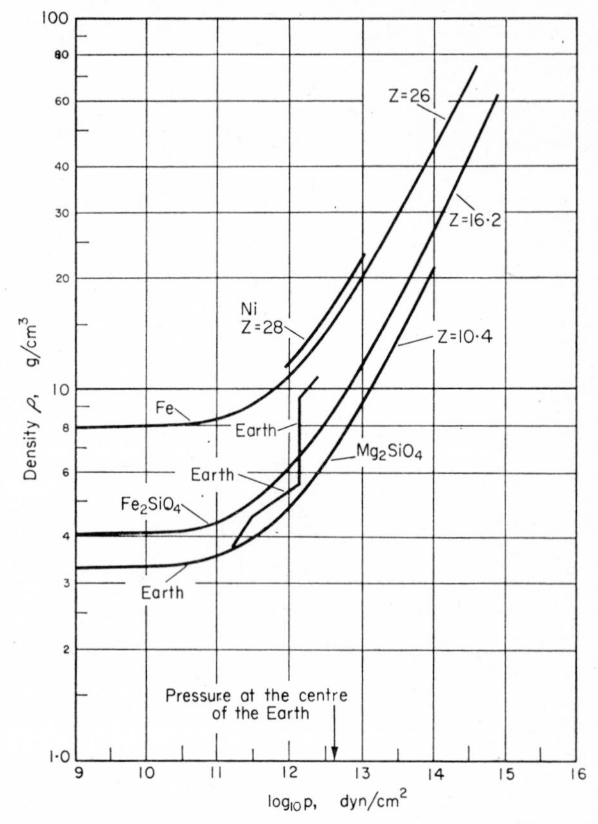

FIG. 3.4. The pressure–density relation for the Earth in relation to the interpolated curves for fayalite (Fe_2SiO_4), forsterite (Mg_2SiO_4), iron and nickel at $T = 0$°K (after L. Knopoff and R. J. Uffen).

Knopoff and Uffen estimate, however, that providing the temperature at the core boundary does not exceed 10,000° the errors should be less than 8 per cent. For the temperature range 0–5300°K the representative atomic number* corresponding to

* The representative atomic number of a compound is defined as the atomic number of a hypothetical pure element which has the same pressure–density relationship as the compound.

the base of the mantle lies between 12·5 and 13·5 with the corresponding range of composition for an olivine mantle of from 47 to 63 per cent Mg_2SiO_4. The representative atomic number corresponding to the outer part of the core was found to be 22, which is intermediate between that of iron and the silicates. It is 4 units less than that for iron and 6 less than that for nickel. Provided that there are no phase transitions, a core with an atomic number of 22, composed of iron, fayalite and forsterite, would have an iron content of nearly 90 per cent.

The composition of the outer core is as yet by no means settled. Most probably it consists mainly of iron alloyed with lighter materials. The inner core is probably composed of material whose mean atomic number is at least equal to that of iron with the possibility of the addition of some material of higher atomic number. Additional evidence on the composition of the outer core is the existence of iron meteorites and the fact that iron would probably be liquid at the temperatures and pressures of the core while silicate materials would be solid (see Section 4.2). G. J. F. MacDonald and L. Knopoff (1958) have shown that a core of composition $(Fe, Ni)_{1\cdot 6}Si$ is consistent with an average composition of the Earth equal to that of chondritic meteorites, provided that the mantle is made up of an upper layer having the composition of eclogite and a lower layer that of dunite. The representative atomic number of $(Fe_{1\cdot 6})Si$ is 22·8. Thus the composition of the core deduced from equation of state arguments is consistent with a chondritic model of the Earth. The estimated electrical conductivity of $(Fe, Ni)_{1\cdot 6}Si$ is also not inconsistent with that required for a dynamo theory of the Earth's magnetic field (see Chapter 6).

References

ALTSHULER, L. V., KRUPNIKOV, K. K., LEDENEV, B. N., ZHUCHIKHIN, V. I. and BRAZHNIK, M. I. Dynamic compressibility and equation of state of iron under high pressure. *J. Exptl. Theoret. Phys. (U.S.S.R.)* **34**, 874–85 (1958); *Soviet Phys. JETP* **34**, 606–14 (1958).

ALTSHULER, L. V., KRUPNIKOV, K. K. and BRAZHNIK, M. I. Dynamic compressibility of metals under pressures from 400,000 to 4,000,000 atmospheres. *J. Exptl. Theoret. Phys.* (*U.S.S.R.*) **34**, 886–93 (1958); *Soviet Phys. JETP* **34**, 614–19 (1958).

BRIDGMAN, P. W. The compression of twenty-one halogen compounds and eleven other simple substances to 100,000 kg/cm^2. *Proc. Am. Acad. Arts Sci.* **76**, 1–7 (1945).

BRIDGMAN, P. W. The compression of 39 substances to 100,000 kg/cm^2. *Proc. Am. Acad. Arts Sci.* **76**, 55–70 (1948).

BULLARD, E. C. The density within the Earth. *Verhandel Ned. Geol. Mijnbouwk Genoot.* (*Geol. Ser.*) **18**, 23–41 (1957).

BULLEN, K. E. Compressibility-pressure hypothesis and the Earth's interior. *Mon. Not. Roy. Astr. Soc. Geophys. Suppl.* **5**, 355–68 (1949).

BULLEN, K. E. An Earth model based on a compressibility-pressure hypothesis. *Mon. Not. Roy. Astr. Soc. Geophys. Suppl.* **6**, 50–59 (1950).

KNOPOFF, L. and UFFEN, R. J. The densities of compounds at high pressures and the state of the Earth's interior. *J. Geophys. Res.* **59**, 471–84 (1954).

KNOPOFF, L. and MACDONALD, G. J. F. An equation of state for the core of the Earth. *Geophys. J.* **3**, 68–77 (1960).

MACDONALD, G. J. F. and KNOPOFF, L. On the chemical composition of the outer core. *Geophys. J.* **1**, 284–97 (1958).

MCQUEEN, R. G. and MARSH, S. P. Equation of state for nineteen metallic elements from shock-wave measurements to two megabars. *J. Appl. Phys.* **31**, 1253–69 (1960).

RAMSEY, W. H. On the constitution of the terrestrial planets. *Mon. Not. Roy. Astr. Soc.* **108**, 406–13 (1948).

UREY, H. C. *The Planets.* Yale Univ. Press (1952).

See also

BULLEN, K. E. Physical properties of the Earth's core. *Ann. Geophys.* **11**, 53–64 (1955).

BULLEN, K. E. Seismology and the broad structure of the Earth's interior. *Physics and Chemistry of the Earth*, Vol. 1, pp. 68–93. Pergamon Press (1956).

CHAPTER 4

Thermal History of the Core

4.1 Temperatures in the Primitive Earth

So far the origin of the Earth and the solar system has not been discussed. These questions, which are still extremely controversial, cannot be avoided in any detailed study of the physics of the Earth's interior. Conditions at depth within the Earth must depend in part on initial conditions, i.e. on the conditions under which the Earth came into being.

All thermal histories of the Earth are based on either a hot or on a cold origin; should a cold origin lead at some later stage in the evolution of the Earth to a molten state, the subsequent thermal histories, whatever the initial origin, would be the same. A cold origin is usually preferred now, and quite low initial temperatures (well below the melting point of the silicate rocks that form the mantle) are generally assumed. Estimates of the time of accretion are of the order of 10^8 years. Possible heat sources that could raise the temperature of the Earth during this period of accretion are the radioactive decay of both long-lived and short-lived isotopes, chemical reactions, and the conversion of kinetic energy into thermal energy. Because of the comparatively short time of accretion, the temperature increase due to the radioactive decay of long-lived isotopes is small, of the order of 150°C (G. J. F. MacDonald, 1959). Short-lived radioactive isotopes could have contributed to the initial heat of the Earth if the time between the formation of the elements and the aggregation of the Earth was short compared with the half lives of the isotopes. The important short-lived isotopes are U^{236}, Sm^{146}, Po^{244} and Cm^{247}, all of which have half lives sufficiently long to have heated up the Earth during the 10^7 to 10^8 years after the initial formation. MacDonald (1959) estimates that if all this heat was retained by the Earth, a temperature increase of the order of 2000–3000°C may be possible.

The temperature of the material within the aggregating Earth will also increase because of adiabatic compression. Although data (particularly on the variation of the coefficient of thermal expansion) are rather uncertain, a rise in temperature from this source of several hundred degrees seems likely. However, the largest source of available energy is the potential energy due to the mutual gravitational attraction of the particles of the dust cloud. This energy, upon aggregation, is either converted into internal energy or radiated away. It is difficult to estimate the total contribution from this source because of the uncertainty of the physical process of accretion. The result depends quite critically on the temperature attained at the surface of the aggregating Earth and on the transparency of the surrounding atmosphere to radiation.

Comparatively low surface temperatures (of the order of a few hundred degrees) have been predicted, mainly because the atmosphere of the primitive Earth was assumed transparent so that the thermal energy of the impinging particles was immediately re-radiated into space. A. E. Ringwood (1960) has argued, however, that during these early years, the primitive Earth will have a large reducing atmosphere. In the presence of these reducing agents (chiefly carbon and methane), the accreting material will be reduced to metallic alloys—principally of iron, nickel and silicon. The outer regions of the Earth will thus be metal rich and dense (referred to zero pressure) compared to the interior. Such a state is gravitationally unstable, and convective overturn will follow leading to a sinking of the metal rich outer region into the centre. This would release further heat due to the energy of gravitational rearrangement. Ringwood believes the whole process is likely to be catastrophic, since the overturn will be accelerated as the initial temperature rises.

Professor H. C. Urey (1962), on the other hand, has put forward convincing evidence that the Earth, accumulating at low temperatures, has at no time in its history become completely molten. The Earth has lost most of its hydrogen, helium and other volatiles, and this must have taken place at low temperatures since otherwise many fairly volatile elements such as mercury, arsenic, cadmium and zinc would have been lost as well, which is not the case. Urey thus argues that any heat must have been lost before this separation and thus was not available for producing high temperatures in an accumulating Earth. Radioactive heating would raise the temperature until viscosity was sufficiently low and convection would then occur. This would dissipate the heat, and no general and complete melting of the Earth is likely to have taken place.

4.2 The Physical State of the Inner Core

Of all the problems connected with the properties of the Earth's interior, its thermal properties are the least precisely known. The reasons for this are threefold. In the first place there is the uncertainty of the chemical and thermodynamic relations which exist at the high pressures deep within the Earth (including possible phase changes); secondly there is the unknown concentration and distribution of radioactivity, and thirdly the present thermal regime is determined to a very large extent by the initial temperature, i.e. by the conditions under which the Earth was formed.

Some results about the inner core can be deduced from purely qualitative reasoning however. The question of whether the Earth ever became completely molten has already been discussed. We know that the mantle of the Earth is solid and the outer core liquid and evidence has already been presented which indicates that the inner core may be solid. If this is so and the Earth ever became completely molten, we are faced with the problem of explaining how the Earth could have cooled leaving the mantle and inner core solid and at the same time leaving the outer part of the core liquid. Assuming the core to consist mainly of iron and the mantle of silicates, J. A. Jacobs (1953) has given the following physical explanation of this point. At the boundary between the mantle and the core there must be a discontinuity in the melting point depth curve (since we have different materials on either side of the boundary), although the actual temperature must be continuous across the boundary. The form of this discontinuity could, mathematically, take any of the three cases shown in Fig. 4.1. Case (i) in which the melting point curve in

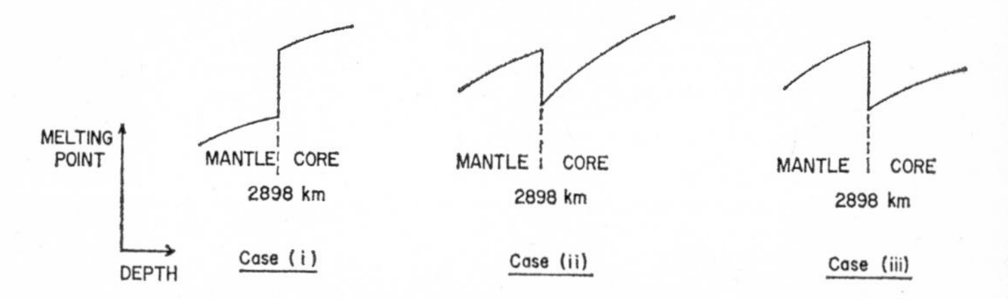

FIG. 4.1. Three possible forms of the melting point depth curve near the core–mantle boundary.

the core is always above that in the mantle is impossible—the actual temperature curve must be below the melting point curve in the mantle, above the melting point curve in the core and yet be continuous across the boundary. Cases (ii) and (iii) are both possible and indeed it is this difference between them which decides whether the inner core is solid or liquid. In case (iii) the melting point curve in the core never rises above the value of the melting point in the mantle at the core boundary, whilst in case (ii) it exceeds this value in part of the core.

Considering first case (ii), the melting point curve will be of the general shape shown in Fig. 4.2. If the Earth cooled from a molten state, there would be strong convection currents and rapid cooling of the surface and the temperature gradient would be essentially adiabatic. Solidification would commence at that depth at which the curve representing the adiabatic temperature first intersected the curve representing the melting point temperature. If as in

Fig. 4.2 the melting point depth curve reaches a value at the centre of the Earth above that at A, the boundary between the core and mantle, then solidification would commence at the centre of the Earth. A solid inner core would continue to grow until a curve

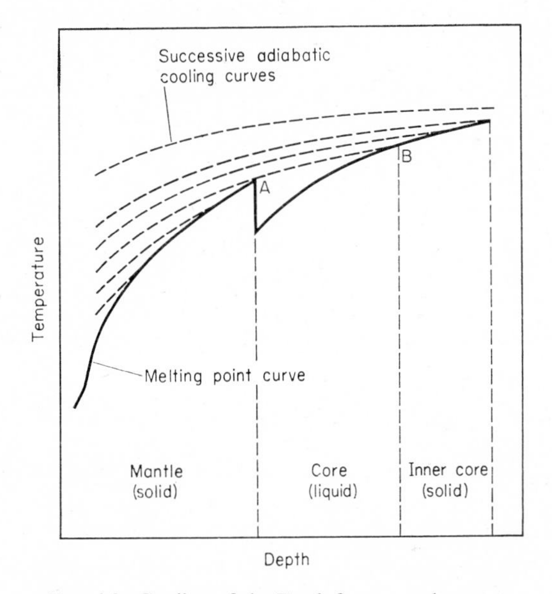

FIG. 4.2. Cooling of the Earth from a molten state.

representing the adiabatic temperature intersected the melting point curve twice, once at A and again at B. As the Earth continued to cool, the mantle would solidify from the bottom upwards, trapping a liquid layer between A and B. The mantle would cool at a relatively rapid rate, leaving this liquid layer essentially at its original temperature, insulated above by a rapidly thickening shell of silicates and below by the already solid (iron) inner core.

In the above argument no specific values of the temperature are given and the behaviour of both the adiabatic and melting point curves need not be known exactly. If they vary qualitatively as shown in Fig. 4.2, then the above argument gives a physical explanation of how the Earth could cool from a molten state, leaving the outer core liquid and the inner core solid. If the melting point depth curve in the core never rises above its value in the mantle at A (case (iii), Fig. 4.1), then similar reasoning shows that as the Earth cooled from a molten state the entire core would be left liquid.

The critical point in these arguments is the magnitude of the discontinuity in the melting point depth curve at the core–mantle boundary. Most evidence (see below) indicates that the melting point at A, the bottom of the mantle, is considerably greater than that at C, the centre of the Earth, i.e. that case (iii), Fig. 4.1, is most likely to occur. In this case solidification of a molten Earth would commence at the core–mantle boundary, and a much greater amount of cooling through the mantle than is generally accepted would be required before the temperature could fall below the melting point curve of the inner core, and so finally leave it solid. Two lines of evidence show that this is a possibility. A. E. Ringwood (1960) has pointed out that in the cooling of a molten Earth it is possible that the crystal mush at the bottom of the mantle may undergo an independent convection, whilst the overlying liquid is still crystallizing and convecting. This would greatly reduce the temperature throughout the solidified zone. When the mantle had completely solidified, the only mechanism for further cooling that has been considered until quite recently, is conduction—which is an extremely slow process. However, a number of workers have examined in detail the possibility of radiative heat transfer in the mantle following the suggestion of F. W. Preston in 1956. Although differing in orders of magnitude, all investigations indicate that the effect of radiative transfer is to increase considerably the effective thermal conductivity of the mantle. Moreover, although the ordinary molecular conductivity will probably decrease in the outer layers of the Earth due to the increase in temperature, R. J. Uffen (1952) has shown that in the deeper parts of the mantle, the effect of pressure may dominate that of temperature and cause a considerable increase in the thermal conductivity. E. A. Lubimova (1960) has considered one further process of heat transfer in the Earth's mantle—energy transfer by the excited states of atoms. At high temperatures, the contribution from the higher excitation energies increases (in normal conditions practically only the lowest excitation energy is important), and thus the rate of the excitation thermal conductivity must increase. Thus all the recent evidence is for a considerable increase in the effective conductivity with depth, and had the Earth undergone considerable cooling by mantle convection in its early history, it is more than possible that it could have cooled further throughout geologic time and formed a solid inner core. The inner core would thus have been slowly growing.

The arguments presented above depend on the behaviour of the melting point and adiabatic gradients. In the last few years a number of estimates have been made of the variation with depth of the melting point of the materials that compose the Earth. Using solid-state theory and seismic data, R. J. Uffen (1952) estimated the melting-point–depth curve in the mantle, while F. E. Simon (1953) used a semi-empirical equation to estimate the melting point of iron and hence obtained melting-point–depth curves in the core. Sir Edward Bullard (1954) has also used Simon's equation to estimate melting points in both the mantle and the core. A feature of all these estimates is much lower values for the melting point in the core than in the mantle. Recent experimental work by H. M. Strong (1959) on the fusion of iron up to a pressure of 96,000 atm, also leads by extrapolation to comparatively low values for the melting point in the core.

V. N. Zharkov (1959) has estimated the melting point of the mantle on the theory that fusion is obtained when the density of thermal defects in the crystalline solid reaches a certain critical value. At depths greater than 1000 km, his results are about 1000°C lower than those obtained by Uffen. On the other hand, his estimates of the melting point of iron at high pressures are greater than those obtained by Simon, his value of the melting point of iron at the core–mantle boundary being approximately the same as that of the mantle at that depth. He thus finds an almost smooth variation of the melting-point–depth curve throughout the entire Earth. J. J. Gilvarry (1956) under certain assumptions, has given a theoretical justification for Simon's equation, and obtained fusion temperatures for both the mantle and the core. Unlike other results, he obtains melting temperatures for the inner core in excess of those in the mantle (1957). On the basis of Gilvarry's curves, the explanation put forward by Jacobs (1953) for the formation of a solid inner core as the Earth cooled from a completely molten state still holds. However, if the Earth had a cold origin and never became completely molten which seems more likely, a solid inner core and liquid outer core are still possible. A rising temperature curve at depth in the interior (due to compression and radioactivity) could eventually meet the melting-point curve at the core–mantle boundary and melting would thence progress inwards leaving a solid inner core.

Assuming the existence of a solid inner core, it is possible to estimate the actual temperature at the core–mantle boundary using values that have been given for the melting point and adiabatic temperature curves for the core. Since the boundary B (see Fig. 4.2) of the inner core is the point of transition between the liquid and solid state in the core, the melting point at B must be the actual temperature there. Hence by drawing the adiabat through B, the actual temperature in the outer liquid core can be estimated. Using Simon's value of 3900°K for the melting temperature at B, Jacobs (1954) obtained a value of 3600°K for the actual temperature at A, the core–mantle boundary. Such arguments, however, can be dangerous since it is all too easy to forget the assumptions on which they rest. Estimates of "actual" temperatures in the above case were based, among other things, on the hypothesis that the inner core is solid—it is all too easy to argue in a vicious circle and use such "actual" temperatures to prove that the inner core is solid!

References

BULLARD, E. C. The interior of the Earth. *The Earth as a Planet*, pp. 57–137. Univ. Chicago Press (1954).

GILVARRY, J. J. Equation of the fusion curve. *Phys. Rev.* **102**, 325–31 (1956).

GILVARRY, J. J. Temperatures in the Earth's interior. *J. Atmos. Terr. Phys.* **1**, 84–95 (1957).

JACOBS, J. A. The Earth's inner core. *Nature* **172**, 297 (1953).

JACOBS, J. A. Temperature distribution within the Earth's core. *Nature* **173**, 258 (1954).

LUBIMOVA, E. A. On the process of heat transfer in Earth's mantle. *Ann. di Geofis.* **14**, 65–78 (1961).

MACDONALD, G. J. F. Calculations on the thermal history of the Earth. *J. Geophys. Res.* **64**, 1967–2000 (1959).

PRESTON, F. W. Thermal conductivity in the depths of the Earth. *Amer. J. Sci.* **254**, 754–57 (1956).

RINGWOOD, A. E. Some aspects of the thermal evolution of the Earth. *Geochim. et Cosmochim. Acta* **20**, 241–49 (1960).

SIMON, F. E. The melting of iron at high pressures. *Nature* **172**, 746 (1953).

STRONG, H. M. The experimental fusion curve of iron to 96,000 atmospheres. *J. Geophys. Res.* **64**, 653–9 (1959).

UFFEN, R. J. A study of the thermal state of the Earth's interior. Ph.D. Thesis, Univ. Western Ontario, London, Canada (1952).

UFFEN, R. J. A method of estimating the melting point gradient in the Earth's mantle. *Trans. Amer. Geophys. Union* **33**, 893–6 (1952).

UREY, H. C. Evidence regarding the origin of the Earth. *Geochim. et Cosmochim. Acta* **26**, 1–13 (1962).

ZHARKOV, V. N. The fusion temperature of the Earth's mantle and the fusion temperature of iron under high pressures. *Bull. (Izv.) Acad. Sci. U.S.S.R. Geophys. Ser.*, 465–70 (1959). (English ed. 1960 pp. 315–20.)

See also

VERHOOGEN, J. Temperatures within the Earth. *Physics and Chemistry of the Earth*, Vol. 1, pp. 17–43. Pergamon Press (1956).

CHAPTER 5

The Earth's Magnetic Field

5.1 Introduction

The Earth's magnetic field is the subject of one of the earliest scientific treatises ever written, *De Magnete*, which was published in 1600 by William Gilbert. The discovery of the directive property of a magnetized needle in the Earth's field and the invention of the mariner's compass is obscure. Claims go back many years and it has been contended that the principles were known to the Chinese some 4500 years ago. It is difficult to substantiate these very early claims which are based on the mythological history of China, and it seems more probable that the earliest reliable evidence of Chinese knowledge is late in the eleventh century. The interpretation of the ancient Greek and Roman literature on this subject is also open to some doubt. The earliest mention in European literature is ascribed to a monk, Alexander Neckham (1157–1217). In spite of this early interest in and practical utilization of the Earth's magnetic field, its origin is still uncertain, and has been the subject of much controversy. At its strongest near the poles the Earth's magnetic field is several hundred times weaker than that between the poles of a toy horseshoe magnet—being less than a gauss (Γ).* Thus in geomagnetism we are measuring extremely small magnetic fields and a more convenient unit is the gamma (γ), defined as $10^{-5}\,\Gamma$.

---

* Strictly speaking the unit of magnetic field strength is the oersted, the gauss being the unit of magnetic induction. The distinction is somewhat pedantic in geophysical applications since the permeability of air is virtually unity in cgs units. The traditional unit used in geomagnetism, the gauss, has been retained in this book.

In contrast the fields of industrial magnets may be thousands of oersteds, and the fields of magnets in cyclotrons and other large high-energy accelerators are of the order of 10,000 oersteds; fields of nearly a million oersteds have been produced in the laboratory. Although the Earth's field is thus by comparison a weak field, it occupies a very large volume and since the energy of a magnetic field is proportional to the volume, the Earth's field plays an important role in extra-terrestrial relationships. It screens the Earth's equatorial regions from cosmic rays with energies of a few tens of billions electron volts and traps charged particles in the Van Allen radiation belts, and is distinguishable from the background field of interplanetary space out to distances of 10–13 Earth radii.

For at least 300 years before Gilbert's famous treatise in 1600 it had been noticed that a suspended magnet did not everywhere point exactly in the direction of geographical or true north. This was at first incorrectly attributed to non-uniform properties of the lodestone which was used to magnetize the needle or to the method of magnetization, but it was gradually appreciated (certainly not later than 1450) that this divergence was a universal phenomenon. In a magnetic compass the needle is weighted so that it will swing in a horizontal plane and its deviation from geographical north is called the declination D. D is reckoned positive or negative according as the deviation is east or west of geographical north. The vertical plane through the magnetic force $\mathbf{F}$ (or its horizontal component H) is called the magnetic meridian. Thus the declination at any point P is the angle between the magnetic meridian and the geographical meridian through P.

It was also noticed early on that a non-magnetic needle which is balanced horizontally on a pivot, becomes inclined to the vertical when magnetized. Over most of the northern hemisphere the north-seeking end of the needle will dip downwards, the angle it makes with the horizontal being called the magnetic dip or inclination I. Over most of the southern hemisphere, the north-seeking end of the needle points upwards and the inclination, I, is considered negative. The total intensity $\boldsymbol{F}$, the declination D and the inclination I completely define the magnetic field at any point, although other components are often used. The horizontal and vertical components of $\boldsymbol{F}$ are denoted by H and Z. H is always considered positive, whatever its direction, while Z is reckoned positive downwards and thus has the same sign convention as I. H is further resolved into two components X and Y. X is the component along the geographical meridian and is reckoned positive if northward; Y is the orthogonal component and is reckoned positive if eastward. Figure 5.1 illustrates these different

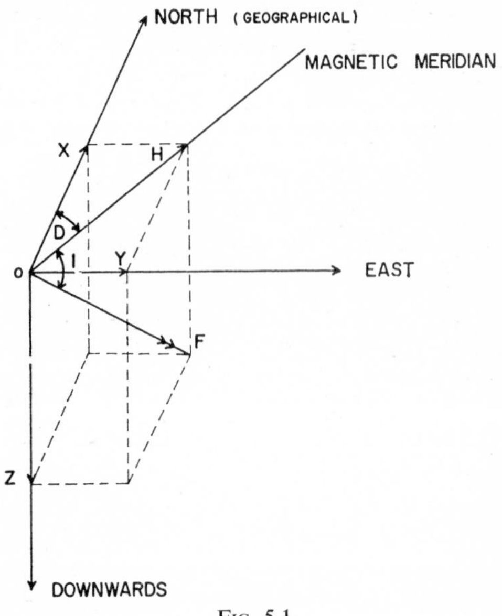

FIG. 5.1.

magnetic elements. They are simply related to one another by th following equations,

$$H = F\cos I, \qquad Z = F\sin I, \qquad \tan I = Z/H \tag{5.1}$$
$$X = H\cos D, \qquad Y = H\sin D, \qquad \tan D = Y/X \tag{5.2}$$
$$F^2 = H^2 + Z^2 = X^2 + Y^2 + Z^2 \tag{5.3}$$

The variation of the magnetic field over the Earth's surface is best illustrated by isomagnetic charts, i.e. maps on which lines are drawn through points at which a given magnetic element has the same value. The earliest geomagnetic chart was drawn by Edmund Halley for the year 1700 and was published in 1701. It shows contours representing lines of equal declination (isogonics) for both the North and South Atlantic oceans. The first chart showing lines of equal inclination (isoclinics) for the whole Earth was published in 1768 by Johann Carl Wilcke. Contours of equal intensity in any of the elements X, Y, Z, H or F are called isodynamics. Figures 5.2 and 5.3 are world maps showing isogonics and isodynamics (for the element H) for the year 1955. It is remarkable that a phenomenon (the Earth's magnetic field) whose origin lies within the Earth should show so little relation to the broad features of geography and geology. The isomagnetics cross from continents to oceans without disturbance and show no obvious relation to the great belts of folding or to the pattern of submarine ridges. In this the magnetic field is in striking contrast to the Earth's gravitational field and to the distribution of earthquake epicentres, both of which are closely related to the major features of the Earth's surface.

H. Gellibrand (1635) discovered that the magnetic declination changed with time. He based his conclusions on the following observations which were made in London.

TABLE 5.1.

Date	Observer	Declination
1580, Oct. 16	William Burrow	11·3° E
1622, June 13	Edmund Gunter	6·0° E
1634, June 16	Henry Gellibrand	4·1° E

This change in the magnetic field with time is called the secular variation and is observed in all magnetic elements. If successive annual mean values of a magnetic element are obtained for a

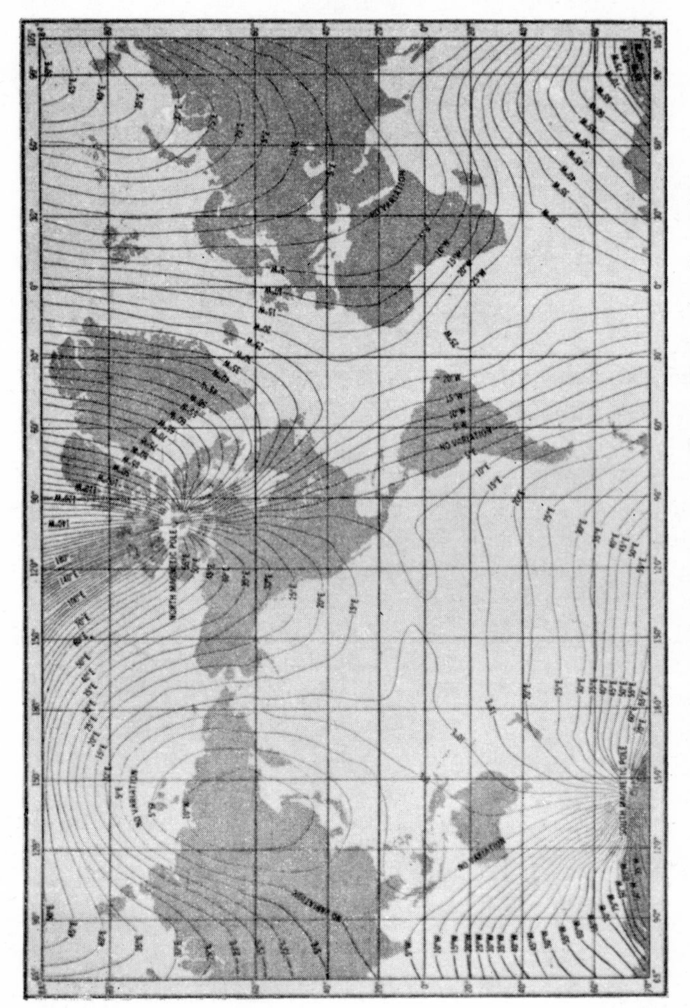

FIG. 5.2. World map showing contours of equal declination (isogonics) for 1955. Courtesy of *Encyclopaedia Britannica*. (Map based on U.S.

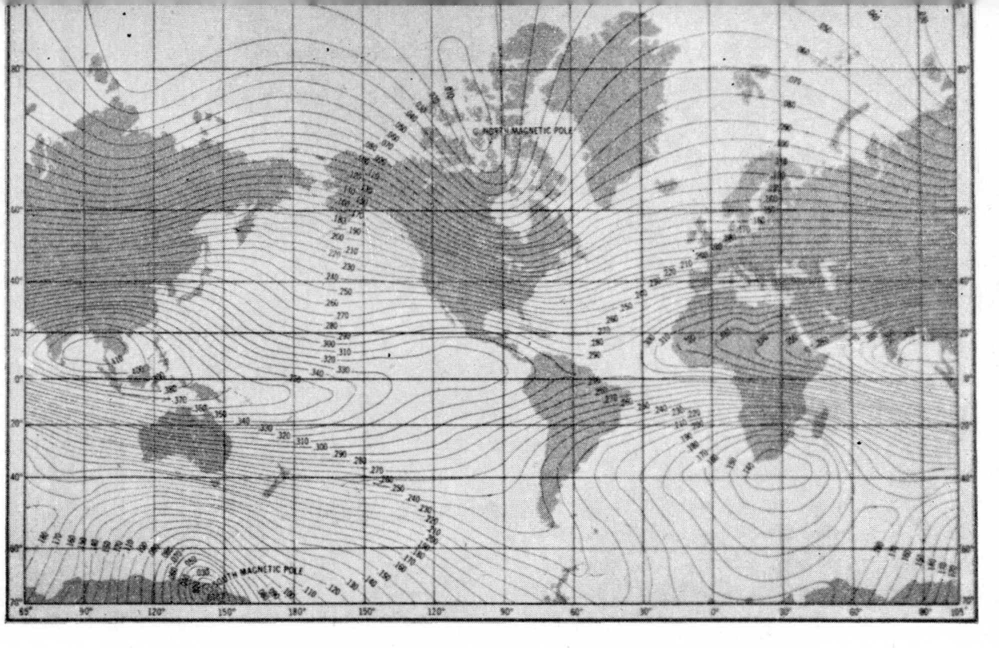

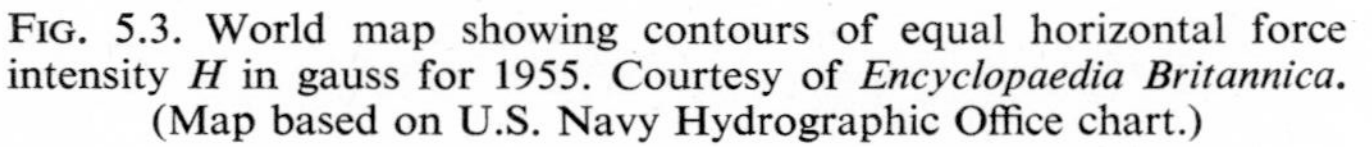

FIG. 5.3. World map showing contours of equal horizontal force intensity H in gauss for 1955. Courtesy of *Encyclopaedia Britannica*. (Map based on U.S. Navy Hydrographic Office chart.)

particular station, it is found that the changes are in the same sense over a long period of time, although the rate of change is not usually constant. Over a period of a hundred years or so this change may be considerable. Thus the H component at Cape Town has decreased by 21 per cent in the hundred years following the first observations in 1843. Figure 5.4 shows the changes in declination and inclination at London and Paris since about 1600. A compass needle at London was $11\frac{1}{2}$°E of true north in 1580 and $24\frac{1}{4}$°W of true north in 1819, a change of almost 36° in 240 years. The curves in Fig. 5.4 suggest that there might be a

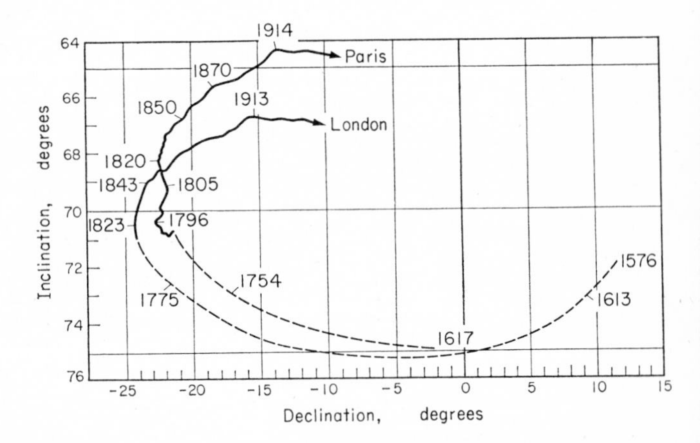

FIG. 5.4. Secular variation of the direction of the geomagnetic field in Paris and London (after Gaibar-Puertas).

cyclic variation. However, the changes at other stations are different and it is very doubtful whether there is any significant periodicity in the secular variation at any station. Lines of equal secular change (isopors) in an element form sets of ovals centring on points of local maximum change (isoporic foci). Figures 5.5 and 5.6 show the secular change in Z for the years 1922.5 and

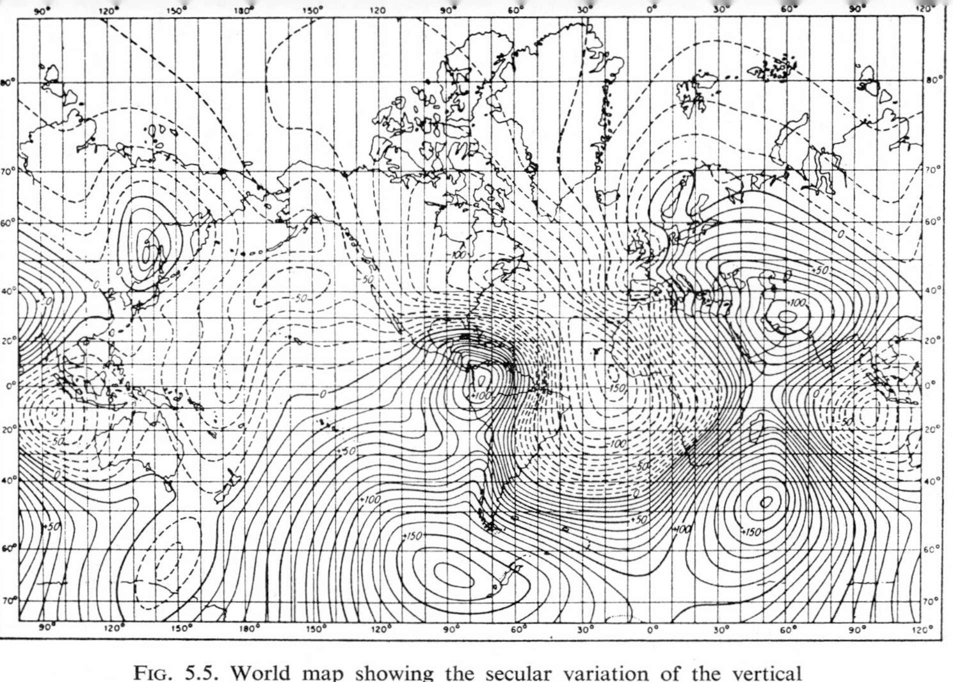

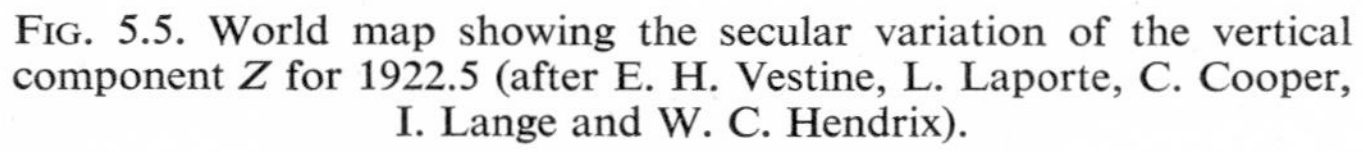

FIG. 5.5. World map showing the secular variation of the vertical component Z for 1922.5 (after E. H. Vestine, L. Laporte, C. Cooper, I. Lange and W. C. Hendrix).

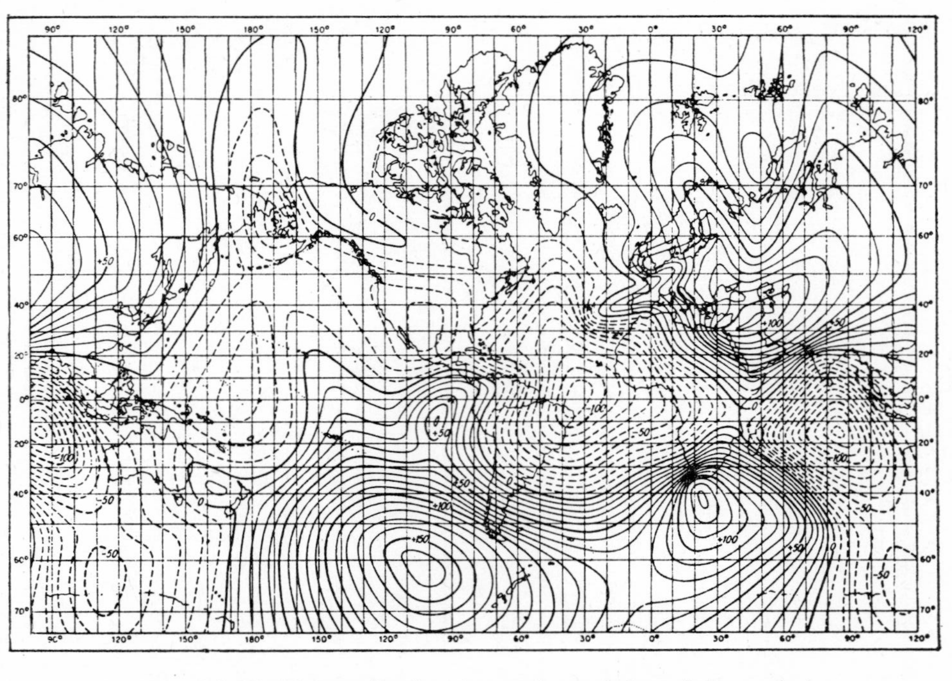

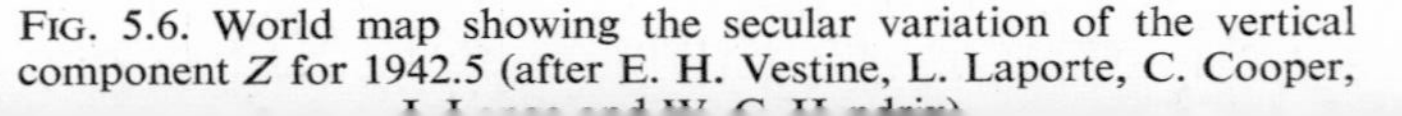
FIG. 5.6. World map showing the secular variation of the vertical component Z for 1942.5 (after E. H. Vestine, L. Laporte, C. Cooper,

1942.5. It is clear that considerable changes take place in the general distribution of isopors even within 20 years. The secular variation is a regional rather than a planetary phenomenon and is anomalously large and complicated over and around the Antarctic. In an area about 1000 km in linear extent between South Africa and Antarctica, the secular change in the total intensity is at present -220γ/year which is about 18 times as large as the average rate. The secular change is particularly remarkable in the area of East Antarctica where at present there are two isoporic foci, one positive and the other negative, the intensities being $+200\gamma$/year and -100γ/year.

Much effort was expended early this century to extend magnetic measurements over the oceans and unexplored regions of the world in an attempt to obtain a more precise description of the Earth's field and secular variation. Much of the initiative came from the Carnegie Institute of Washington, whose non-magnetic ship the *Carnegie* was unfortunately destroyed by fire in Apia harbour, Samoa, in 1929. Since then few other cruises by non-magnetic ships have been made until the Soviet Union fitted out their non-magnetic ship *Zarya* during the International Geophysical Year (I.G.Y.). E. H. Vestine and his colleagues (1947) have carried out an exhaustive analysis of the main geomagnetic field and its secular variation between the years 1912 and 1942—in particular they have published detailed charts and tables for the four epochs 1912.5, 1922.5, 1932.5 and 1942.5. World maps for the different magnetic elements and their secular variation have been published for the epoch 1955.0 by the U.S. Navy Hydrographic Office and will be re-issued at fixed intervals in the future.

The programme for geomagnetism during the I.G.Y. called for a world wide survey of the Earth's magnetic field. However, the I.G.Y. was timed to coincide with a period of maximum solar activity when magnetic storms and other disturbances were at a peak, and many phases of the magnetic survey were delayed until the field could be measured with a minimum of interference from solar induced disturbances. The solar cycle has a period of approximately 11 years and present plans call for the major effort in the World Magnetic Survey (W.M.S.) to be made during the International Year of the Quiet Sun (I.Q.S.Y.) which extends from April 1964 to December 1965.

On land the survey can be carried out by the network of permanent observatories supplemented by a number of temporary stations. The values of the field strength can be determined to an accuracy of 1 γ and declination and inclination to 0·2 min of arc. Since more than 70 per cent of the Earth's surface is covered by water, many measurements at sea are necessary in order to provide sufficient coverage for accurate mapping. Ship measurements can obtain values of the magnetic elements to an accuracy of about 1 part in 300. However, much time is necessary to cover the vast oceanic regions and there is at present only one non-magnetic ship (the *Zarya*) which can make vector measurements of the magnet field. Many oceanographic research vessels, however, use a magnetometer mounted in a fish and towed sufficiently far behind the ship so that the magnetic field of the ship does not influence the readings. Only total field strength F is generally obtained in this way.

Measurements have been made by aircraft for some time now and have proved a valuable tool in geophysical exploration for mineral deposits. The detail is too fine for mapping purposes, however, since the flights have been carried out at low altitudes in a search for local anomalies. Originally airborne surveys measured only the total field, but it has now proved possible to install 3 component magnetometers which will measure F, D and I. Canadian scientists have surveyed much of Canada from the air and have also made crossings of the Atlantic and Pacific oceans. Total intensity can be measured with an accuracy of $\pm 15\gamma$ and angular directions to within $\pm 0{\cdot}1°$ of arc. A major programme of airborne surveys (Project Magnet) is being carried out by the U.S. Navy Hydrographic Office—several flights have already been completed in the Antarctic.

5.2 The Field of a Uniformly Magnetized Sphere

William Gilbert published in 1600 the results of his experimental studies in magnetism. He investigated the variation in direction of the magnetic force over the surface of a piece of the naturally magnetized mineral lodestone which he had cut in the shape of a sphere. He found that the variation of the inclination was in agreement with what was then known about the Earth's magnetic field, and he came to the conclusion that the Earth behaved substantially as a uniformly magnetized sphere, its magnetic field being due to causes within the Earth and not from any external agency as was supposed at that time. Since 1600 the direction and intensity of the Earth's magnetic field has been measured at many widely scattered points over the Earth's surface, although no attempt was made to represent the field mathematically before 1839. In that year Gauss, by a spherical harmonic analysis (see Appendix B), showed that the field of a uniformly magnetized sphere was an excellent first approximation to the Earth's magnetic field. Gauss further analysed the irregular part of the Earth's field, i.e. the difference between the actual observed field and that due to a uniformly magnetized sphere. With the data then available, he showed that both the regular and irregular components of the Earth's field were of internal origin.

Since the north-seeking end of a compass needle is attracted towards the northern regions of the Earth, those regions must have opposite polarity. Consider therefore the field of a uniformly magnetized sphere whose magnetic axis runs north–south, and let P be any external point distant r from the centre O, and θ the angle NOP, i.e. θ is the magnetic co-latitude (see Fig. 5.7). Equations (A.7) and (A.8) give the radial and cross-radial components of force at an external point due to a uniformly magnetized sphere. (See Appendix A. The field of a uniformly magnetized sphere can be represented by a dipole at its centre). Since OP makes an angle $\pi-\theta$ with the direction of the magnetic axis, the magnetic component Z, which is the *inward* radial component of force, is given by

$$Z = \frac{-2M \cos (\pi - \theta)}{r^3} = \frac{2M \cos \theta}{r^3} \tag{5.4}$$

and the magnetic component H by

$$H = \frac{M \sin (\pi - \theta)}{r^3} = \frac{M \sin \theta}{r^3} \tag{5.5}$$

The inclination I is thus given by

$$\tan I = \frac{Z}{H} = 2 \cot \theta \tag{5.6}$$

and the magnitude of the total force F by

$$F = (H^2 + Z^2)^{\frac{1}{2}} = \frac{M}{r^3}(1 + 3 \cos^2 \theta)^{\frac{1}{2}} \tag{5.7}$$

The maximum value of Z on the surface of the sphere ($r = a$) is

$$Z_0 = \frac{2M}{a^3} \tag{5.8}$$

and occurs at the poles. The maximum value of H on the surface is

$$H_0 = \frac{M}{a^3} \tag{5.9}$$

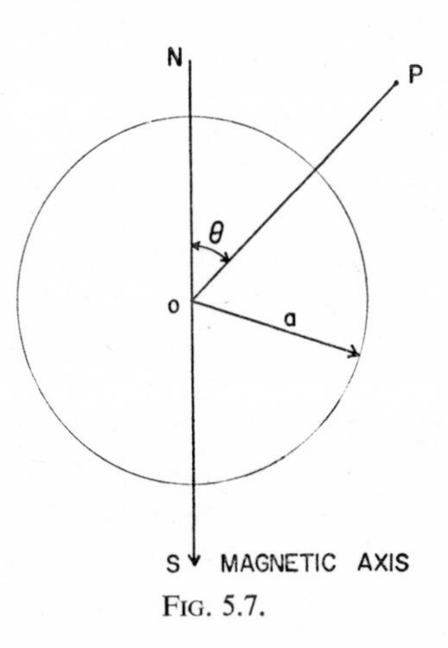

FIG. 5.7.

and occurs at the equator. Thus $Z_0 = 2H_0$ which relationship is approximately true for the Earth's field. (The total intensity F of the Earth's magnetic field is a maximum near the magnetic poles—its value is just over $0{\cdot}6\,\Gamma$ near the northern dip pole and just over $0{\cdot}7\,\Gamma$ near the southern dip pole. Its minimum value of about $0{\cdot}25\,\Gamma$ is near the Tropic of Capricorn off the west coast of South America. In some areas the magnitude of F may exceed $3\,\Gamma$ but this is entirely due to local concentrations of magnetic ore bodies.)

At points on the Earth's surface where the horizontal component of the magnetic field vanishes, a dip needle will rest with its axis vertical. Such points are called dip poles. Two principal poles of this kind are situated near the north and south geographical poles and are called the magnetic north and south poles. Their positions at present are approximately 75°N, 101°W and 67°S, 143°E. They are thus not diametrically opposite, each being about 2300 km from the point antipodal to the other. The magnetic poles must not be confused with the geomagnetic poles which are the points where the axis of the geocentric dipole, which best approximates the Earth's field, meets the surface of the Earth. The geomagnetic poles are situated approximately at 78½°N, 69°W and 78½°S, 111°E and the geomagnetic axis is thus inclined at 11½° to the Earth's geographical axis. If the geocentric dipole field were the total field, the dip poles and geomagnetic poles would of course be the same. A better approximation to the Earth's field can be obtained by displacing the centre of the equivalent dipole by about 300 km towards Indonesia. Vestine (1953) has determined the position of the eccentric dipole from 1830 to 1945 and found a change in longitude of about 0·30° per year.

5.3 The Origin of the Earth's Magnetic Field

A magnetic field observed at the surface of the Earth could be produced by sources inside the Earth, by sources outside the Earth's surface or by electric currents crossing the surface. A spherical harmonic analysis of the observed field (see Appendix B) shows that the source of the field is predominantly internal. Superimposed on this field, however, is a rapidly varying external field giving rise to transient fluctuations. Unlike the secular variation, which is of internal origin, these transient fluctuations produce no large or enduring changes in the Earth's field. They are mostly due to solar effects which disturb the ionosphere and give rise to a number of related upper atmospheric phenomena such as magnetic storms and aurora. Such events are dealt with in other texts in the present series and will not be discussed in this book which is concerned only with the Earth's main field and secular variation.

There has been much speculation as to the cause of the Earth's main field and no completely satisfactory explanation has as yet been given. We will consider all possible ways in which a magnetic field can be produced and see which, if any, are applicable to the Earth's field.

Consider first the possibility of permanent magnetization. The value of H_0 for the regular part of the Earth's field (1945) was $0{\cdot}312\,\Gamma$. Equation (5.9) shows that this corresponds to a magnetic moment $M = H_0 a^3$ and if the field were due to uniform magnetization throughout the Earth, the intensity of magnetization would be given by

$$J = M\Big/\frac{4}{3}\pi a^3 = \frac{3H_0}{4\pi} \simeq 0{\cdot}075\,\Gamma \tag{5.10}$$

The surface layers of the Earth are in general not magnetized to anything like this extent. The situation is much worse than this, however. The temperature gradient in the crust is approximately 30°C/km so that at a depth of about 25 km a temperature of the order of the Curie point for iron, viz. 750°C, is reached. Thus, unless the Curie point increases with increasing pressure, all ferromagnetic substances will have lost their magnetic properties at greater depths. There is no experimental evidence for a large increase in the Curie point with increasing pressure.

Thus in order to account for the observed value of the Earth's magnetic moment, an intensity of magnetization in the Earth's crust of about 6 Γ is necessary which is impossible. Moreover the high rate of change and westward drift of the non-dipole field with velocities up to 20 km/year (see Section 6.3) are difficult to explain as due to geologic processes in the upper mantle or crust.

One of the greatest anomalies in the geomagnetic field is at Kursk, south of Moscow, where along strips 250 km in length, Z is everywhere above normal, reaching values up to 1·9 Γ. The origin of these anomalies is the induced and permanent magnetization of rocks in the Earth's crust. The magnetization of rocks depends on the amount and nature of the iron oxide minerals contained in them, and for most rocks the intensity is less than 10^{-2} Γ. S. A Deel (1945) first pointed out the absence of anomalies on a scale intermediate between local ones (of crustal origin) and those of some thousands of kilometres in extent as are observed on world maps. Presumably therefore there are no appreciable magnetic sources between the Curie point isotherm and the Earth's core. The inner core is probably solid iron and under the high pressures that exist there might be ferromagnetic, and motions in the liquid core could be postulated to provide the electromagnetic induction associated with the secular variation. The intensity of magnetization of the inner core would have to be about 10 Γ. One cannot completely rule out this possibility—one of the difficulties of such a theory is the increasing evidence of the occurrence of reversals of polarity in the geomagnetic field (see Chapter 7).

Consider now the possibility that the Earth's magnetic field arises from the rotation of the Earth. In a ferromagnetic body the magnetic moments of the atoms are associated with their angular momenta. For a rotating body gyroscopic effects cause a partial lining up of the elementary magnetic moments along the axis of rotation. Such a body is uniformly magnetized with an intensity proportional only to its angular velocity and independent of the size of the body. The effect can be observed in the laboratory at high speeds, but the angular velocity of the Earth is so small that the magnetization arising from such causes is negligible—being less than about $10^{-6}\gamma$.

If the Earth is assumed to have either a surface or a volume electric charge, a magnetic field would be produced by its rotation. However, the field produced in this way would appear different to an observer on the Earth and to one in space not sharing the Earth's rotation. If this difficulty is removed by considering the Earth electrostatically neutral—a negative surface charge and a positive volume charge leading to a net dipole moment—electric fields of the order of 10^8 V/cm are required to account for the Earth's magnetic moment. Such potential gradients are many orders of magnitude greater than the dielectric strength of most materials.

W. F. G. Swann (1927) modified the equations of electromagnetism by the introduction of small terms depending upon the acceleration as well as the velocity of electric charges. An alternative approach was made by T. Schlomka (1933) who postulated small deviations from the fundamental laws of electrostatics for the forces between electrons and protons. Such changes in the fundamental laws of physics which would only manifest themselves in rotating bodies of cosmic size have not had much success. However, a further attempt was made by P. M. S. Blackett (1947) to explain the Earth's magnetic field as a new fundamental property of rotating matter on a cosmic scale. He suggested that the dipole moment of massive rotating bodies was proportional to their angular momentum. This was prompted by the discovery by H. W. Babcock (1947) that the star 78 Virginis possessed a magnetic field.

If such a fundamental property did exist it should be possible to detect it by measuring the radial variation of the geomagnetic field with depth. Blackett's theory would imply that the cause of the Earth's field is distributed throughout the bulk of the Earth so that the horizontal field should decrease with depth. Conversely assuming the field to originate in the core, the field should increase with depth as $1/r^3$. No measurements have, of course, been made at any appreciable depth within the Earth, but measurements by S. K. Runcorn *et al.* (1951) in coal mines in England gave no support to Blackett's theory. The discovery of the apparent reversals of stellar magnetic fields and of the Sun's, and the strong possibility of reversals of the Earth's magnetic field (see Chapter 7) has thrown further doubt on the hypothesis.

During the Danish Galathea expedition (1950–1952) measurements were made in the Pacific Ocean of the vertical gradient of the geomagnetic field. Although the existence of large local anomalies over a rugged ocean bottom cast doubt on some of the measurements, most results indicated a definite increase in H and Z with increasing depth with not a single decrease. A further consequence of Blackett's fundamental hypothesis is that a small magnetic field should be produced by a dense body rotating with the Earth. In a long and detailed paper, Blackett (1952) describes the results of a "negative experiment" in which a sensitive astatic magnetometer failed to detect a field of the order of magnitude predicted near dense bodies at rest in the laboratory. Finally any rotational theory can only explain the component of the magnetic moment along the axis of rotation—the transverse component is at present about one-fifth of the axial. We are thus compelled, if a little reluctantly, to abandon any rotational theories as a source of the Earth's main magnetic field. There remains the possibility that electric currents flow in the Earth's interior and set up a magnetic field by induction. This possibility will be discussed in some detail in the next chapter.

References

For an account of the early history of geomagnetism, the reader is referred to A. Crichton Mitchell, Chapters in the history of terrestrial magnetism, *Terr. Mag.* **37**, 105–46 (1932); **42**, 241–80 (1937); **44**, 77–80 (1939). The famous treatise *De Magnete* by William Gilbert, originally published in Latin in 1600 has been translated into English by P. F. Mottelay and published by Dover Publications, Inc. Volume 5 of the collected works of C. F. Gauss contains his two famous memoirs on geomagnetism, *Intensitas vis magneticae terrestris ad mensuram absolutam revocata* (1832) and *Allgemeine Theorie des Erdmagnetismus* (1838).

S. CHAPMAN and J. BARTELS, *Geomagnetism*, Oxford University Press (1940), has been a standard text on geomagnetism for many years. S. CHAPMAN has also published a small book in Methuen's Monographs on Physical Subjects entitled *The Earth's Magnetism* (1951).

BABCOCK, H. W. Zeeman effect in stellar spectra. *Astrophys. J.* **105**, 105–19 (1947).

BLACKETT, P. M. S. The magnetic field of massive rotating bodies. *Nature* **159**, 658–66 (1947).

BLACKETT, P. M. S. A negative experiment relating to magnetism and the Earth's rotation. *Phil. Trans. Roy. Soc. A* **245**, 309–70 (1952).

RUNCORN, S. K., BENSON, A. C., MOORE, A. F. and GRIFFITHS, D. H. Measurements of the variation with depth of the main geomagnetic field. *Phil. Trans. Roy. Soc. A* **244**, 113–51 (1951).

SCHLOMKA, T. Zur physikalischen theorie des Erdmagnetismus. *Z. f. Geophysik* **9**, 99–109 (1933).

SWANN, W. F. G. A generalization of electrodynamics, consistent with restricted relativity and affording a possible explanation of the Earth's magnetic and gravitational fields, and the maintenance of the Earth's charge. *Phil. Mag.* **3**, 1088–1136 (1927).

VESTINE, E. H., LAPORTE, L., COOPER, C., LANGE, I. and HENDRIX, W. C. Description of the Earth's main magnetic field and its secular change, 1905–1945. *Carnegie Inst. Wash. Publ.* No. 578 (1947).

VESTINE, E. H., LANGE, I., LAPORTE, L. and SCOTT, W. E. The geomagnetic field, its description and analysis. *Carnegie Inst. Wash. Publ.* No. 580 (1947).

VESTINE, E. H. On variations of the geomagnetic field, fluid motions and the rate of the Earth's rotation. *J. Geophys. Res.* **58**, 127–45 (1953).

CHAPTER 6

The Dynamo Theory of the Earth's Magnetic Field

6.1 Introduction

Let us assume that there are electric currents flowing in the core of the Earth. These may have started originally by chemical irregularities which separated charges and thus initiated a battery action, generating weak currents. Palaeomagnetic measurements have shown that the Earth's main field has existed throughout geologic time and that its strength has never differed widely from its present value. H. Lamb showed in 1883 that electric currents generated in a sphere of radius R and electrical conductivity σ and left to decay freely would be reduced by electrical dissipation by Joule heating to e^{-1} of their initial strength in a time not longer than $4\sigma\mu R^2/\pi$. This time is of the order of 10^5 years whereas the age of the Earth is more than 4×10^9 years. This decay time is even shorter for the higher harmonics so that the geomagnetic field cannot be a relic of the past, and a mechanism must be found for generating and maintaining electric currents to sustain the Earth's continuing magnetic field. A process that could accomplish this is the familiar action of the dynamo. The dynamo theory of the Earth's magnetic field was due originally to Sir Joseph Larmor who, in 1919, suggested that the magnetic field of the sun might be maintained by a mechanism analogous to that of a self-exciting dynamo. This suggestion was followed up by W. M. Elsasser and Sir Edward Bullard who have carried out independently a considerable amount of research on the problem, and we shall consider in some detail later. We know that the Earth has a fluid core, the main constituent of which is iron.

Thus the core is a good conductor of electricity and a fluid in which motions can take place, i.e. it permits both mechanical motion and the flow of electric current and the interaction of these could generate a self sustaining magnetic field. The secular variation also lends support to a dynamo theory, the variations and changes in the Earth's magnetic field reflecting eddies and changing patterns in the motions in the core. The recognition by W. M. Elsasser (1939) that the short time scale of the secular variation could be understood if the origin of the field was associated with motions in the Earth's core was the starting point of modern theory. Possible causes of these fluid motions will be discussed in Section 6.2.

It has not proved possible to demonstrate the existence of such a dynamo action in the laboratory. If a bowl of mercury some 30 cm in diameter is heated from below, then thermal convection in the mercury will be set up—but no electric currents or magnetism can be detected in the bowl. Such a model experiment fails because electrical processes and mechanical processes do not scale down in the same way (see Appendix D). An electric current in the bowl of mercury would have a decay time of about one-hundredth of a second. The decay time, however, increases as the square of the diameter of the bowl—and an electric current in the Earth's core would last for about 10,000 years before it decayed. This time is more than sufficient for the current and its associated magnetic field to be altered and amplified by motions in the fluid, however slow. The dynamo theory suggests that the magnetic field is ultimately produced and maintained by an induction process, the magnetic energy being drawn from the kinetic energy of the fluid motions in the core. A group of particles moving at different speeds in the fluid may pull laterally on some magnetic lines of force, thus stretching them. In this process of stretching they will gain energy—energy which is taken from the mechanical energy of the moving particles.

Even after the existence of energy sources sufficient to maintain the field has been established, there remains the outstanding problem of sign: i.e. it must also be shown that the inductive reaction to an initial field is regenerative and not degenerative. The problem is extremely complex involving both hydrodynamic and electromagnetic considerations, and a complete solution may not be possible (see Appendix D). The study of the flow of electrically conducting fluids in the presence of magnetic fields is called "magneto-hydrodynamics" or "hydromagnetics" and has become increasingly important in many branches of geophysics in recent years. In the present case the question that has to be answered is—do there exist motions of a simply-connected, symmetrical, homogeneous and isotropic fluid body which will cause the body to act as a self-exciting dynamo and produce a magnetic field in the absence of any sustaining field from an external source? In an engineering dynamo, the coil has the symmetry of a clock face in which the two directions of rotation are not equivalent—it is this very feature which causes the current to flow in the coil in such a direction that it produces a field which reinforces the initial field. A simple body such as a sphere does not have this property—any asymmetry can exist only in the motions. This is the crux of the problem—whether asymmetry of motion is sufficient for dynamo action or whether asymmetry of structure is necessary as well.

Other possible causes of electromotive forces deep within the Earth are thermoelectric and chemical. A likely place for either would be a major contact between dissimilar materials or between the same substances under markedly different physical conditions. The suggestion that thermoelectric currents circulate within the Earth was first put forward by Elsasser in 1939. Thermoelectric electromotive forces are generated whenever two materials with different electrical properties are in contact at points which are at different temperatures. Elsasser proposed that thermoelectric electromotive forces are due to inhomogeneities in the core material which are created and continuously regenerated by turbulent fluid motion. Only a small fraction of the current generated by such a mechanism could be responsible for the surface magnetic field—the greater part producing only a contained field.* S. K. Runcorn (1954) suggested that thermoelectric currents are generated at the core–mantle boundary where there is a contact between two materials with different electrical properties. Temperature differences between different parts of the core–mantle boundary could be due to the eccentricity of the Earth (producing a temperature contrast between the poles and the equator) and/or thermal convection in the core. All thermoelectric currents generated by Runcorn's hypothesis would produce contained fields and further recourse has to be made to inductive interaction between these fields and the fluid flow in the core. Since the thermoelectric power of materials under the conditions prevailing in the core is unknown, it is extremely difficult to make a quantitative assessment of the thermoelectric theory. It seems that rather extreme assumptions are necessary to make any theory satisfactory—either an extreme geometry or extreme and implausible values of some of the physical properties of the material in the core and lower mantle. It also appears that the convective heat flow demanded by the theory is excessive and it is not at all certain that the required temperature differences can be realized.

6.2 Fluid Motions in the Earth's Core

After over a century of study of dynamical meteorology, our knowledge of so readily an observable phenomenon as the general circulation of the atmosphere is still very far from complete. It is not surprising therefore that the complicated problems of the dynamics of the Earth's core are as yet not fully understood. In the Earth's core fluid motions and electric currents will exist simultaneously—and will interact with each other causing rather complex dynamical behaviour. Most naturally occurring fluid

* I.e. a field with lines of force not cutting the Earth's surface. An electric current flow restricted to meridian planes, for example, would produce a purely zonal magnetic field, i.e. a contained field.

motions are due ultimately to the action of gravity. The gravitational potential near the Earth may be divided into two components—that due to the Earth itself and that due to extra-terrestrial sources (the sun and the moon). The possible effects of these two components on motions in the core will be discussed. Considering first gravitational fields of extra-terrestrial origin, the mantle suffers acceleration due to four effects.

(1) *The Bodily Tide of the Earth*

The mantle undergoes a radial tidal oscillation due to the moon's influence on the rotating Earth. It has been estimated that the amplitude of the oscillation at the core–mantle boundary is of the order of 6 cm and W. M. Elsasser (1950) has shown that it would have a negligible effect on any core motions.

(2) *A Gradual Deceleration of the Rate of the Rotation of the Earth due to Tidal Friction in the Oceans*

E. C. Bullard (1949) has shown that electromagnetic coupling between the mantle and core is so strong that the core is compelled to follow the tidal deceleration of the mantle with but little energy exchanged across the core–mantle boundary.

(3) *Precession and Nutation*

The rotational axis of the Earth precesses about a 24° cone with a period of 27,000 years. Nutation is a fluctuation of smaller amplitude and period superposed on this precession. If the core fails to precess with the mantle, strong fluid motions might be induced which would give rise to much stronger magnetic fields than are observed. It is for this reason that Bullard (1949) suggested that precession (and nutation) do not have an appreciable effect on core motions. The mathematical difficulties of a complete discussion of the problem are, however, very great and the conclusions reached are rather unsatisfactory. Precession cannot definitely be ruled out as a major influence on fluid motions in the core.

(4) *Sudden Changes in the Rate of the Earth's Rotation*

This will be discussed in more detail in Section 6.3. It appears that such irregularities in the rotation of the Earth are intimately tied up with the secular variation. However, sudden changes in the length of the day are more likely to be a consequence of, rather than the cause of, fluid motions in the core.

Consider now the Earth's own gravitational field, which could generate motions in the core if there are density inhomogeneities. This could be brought about in two ways. The first, proposed by H. C. Urey (1952) is that the interior of the Earth is not at rest but that gradual chemical differentiation and slow relative displacement of different constituents is taking place all the time, i.e. the core is continually growing at the expense of the mantle. Urey has developed a theory of the origin of the Earth based on this hypothesis—iron in the mantle slowly and continuously "seeping" into the core. The second process is thermal convection which will occur in the core if the transport of heat radially outwards exceeds the heat transport by thermal conduction alone. The material of the liquid (outer) core should be very nearly in a state of chemical equilibrium, uniformity being maintained by the mixing action of the convective motion. Thus it is unlikely that any motion is due to variations in the physical properties of the liquid outer core—rather it must be determined by boundary conditions. Three possible models have been proposed by W. M. Elsasser (1950). The first is one in which heat supplied by the inner core to the outer liquid layer exceeds the amount that can be carried away from there by conduction alone. The second is one in which the heat flow in the mantle adjacent to the core exceeds the purely conductive heat transport in the core itself. The third is a compromise model in which the heat supplied by the inner core is carried to the core–mantle boundary by convection and is then carried away by the mantle.

The conductive heat flow q is calculated from the relation $q = k\tau$, where k is the thermal conductivity and τ the temperature gradient. Following E. C. Bullard (1950) and W. M. Elsasser (1950) we take the value of k in the liquid iron core to be 0·18 cal/cm.sec.deg. Since the core is in a state of convective agitation, the temperature gradient will be the adiabatic. The adiabatic temperature gradient in the core has been estimated by J. A. Jacobs (1953) who obtained the values $\tau = 0{\cdot}30$ deg/km at the boundary between the core and mantle, and $\tau = 0{\cdot}10$ deg/km at the boundary of the inner core. Thus the conductive heat flow at the boundary of the inner core is $1{\cdot}8 \times 10^{-7}$ cal/cm^2.sec, and if the heat flow Q there exceeds this, convection currents will be set up. Let us suppose, for instance, that Q is double this amount, that is, $3{\cdot}6 \times 10^{-7}$ cal/cm^2.sec, and compare this figure with the radioactive heat developed in the Earth's crust. A large part of the heat flow in the crust is due to radioactivity, so that the measured mean heat outflow from the crust, namely $1{\cdot}2 \times 10^{-6}$ cal/cm^2.sec, should be representative of the order of magnitude of the radioactive heat developed in the crust. This is about $3\frac{1}{2}$ times the value of Q. Since the surface area of the inner core is only about 5 per cent of that of the crust, the radioactive content of the inner core need be less than 2 per cent of that in the crust. But even this small amount presents difficulties, since there are geochemical problems in having any radioactivity in the inner core and neutron activation analyses have failed to find any in iron meteorites (G. W. Reed and A. Turkevitch 1956; G. L. Bate and others 1958).

It is instructive therefore to consider Elsasser's second model, particularly in view of the large increase in the estimate of the effective thermal conductivity with depth in the mantle due to radiative heat transfer (see Section 4.2). If the heat carried away in the mantle exceeds that supplied by conduction alone in the core, the second model will hold. Extra heat could be supplied in the mantle by radioactive material distributed throughout the mantle, although such heat sources are not necessary. The convective process will occur if the heat conduction in the mantle is large enough, the net result being an average loss of heat from the core over geologic time. It is extremely difficult to estimate the relative magnitudes of the heat carried away by conduction in the mantle to that supplied by conduction in the core. The temperature gradient in the liquid core will be approximately the adiabatic, whilst that in the adjacent mantle should not be too different. Thus if the (effective) thermal conductivity of the mantle at the core–mantle boundary is greater than that of the core there, this model will hold. This may well be the case since the effective thermal conductivity of the mantle at depth may be as much as a hundred times that of surface rocks because of radiative transfer and thus greater than that in the metal-rich core where the conductivity will be essentially molecular.

The above discussion is not too satisfactory, since we have no means of estimating with any real certainty the total heat generated in the core, and our estimates of such physical parameters as the thermal conductivity of the core may well be out by a factor of 5 or more. We have also neglected the inhibiting effects of viscosity, thermal conductivity, rotation and magnetic fields on convection (see Appendix D).

6.3 The Secular Variation and Westward Drift

Additional evidence that the core of the Earth is fluid, having motions more rapid than would be expected in the plastic flow of a solid, is found in the secular variation of the geomagnetic field. As early as 1692, E. Halley noticed that certain non-axial features of the geomagnetic field drifted westwards (at a rate of about 0·5°/yr). W. Van Bemmelen (1899) constructed isogonic charts at intervals of 50 years covering the period 1550–1750 and found that a region of maximum declination moved from the Baltic to the east Atlantic, a distance of about 25° in 250 years. L. A. Bauer (1895) and V. Carlheim-Gyllenskold (1907) both found indications of a westerly drift in their analyses of the secular variation. W. M. Elsasser (1950) called attention to the westward motion of the point of zero declination in the western hemisphere at the equator—in the past 400 years its rate has been on the average 0·22°/yr. The exhaustive analyses of E. H. Vestine and his colleagues (1947) who produced secular variation charts for the epochs 1912.5, 1922.5, 1932.5 and 1942.5 and the work of C. Gaiber-Puertas (1953) indicate a clear tendency for the isoporic foci to drift westwards. E. C. Bullard *et al.* (1950) carried out a statistical analysis of Vestine's charts and found the mean westward drift between 1912.5 and 1942.5 of the non-dipole field to be 0·18°/year and of the secular variation field 0·32°/year. (The secular change arises not only from the movement of the non-axial components of the main field with respect to the Earth's surface, but also from field variations within the Earth's core—only about one-third of the rate of the secular variation at any one time seems to be due to the westward drift although its contribution causes the more persistent and steady change in the field observed at any one place when results over hundreds of years are considered (F. J. Lowes 1955). That the westward drift of the residual field is only about a third of the secular variation pattern at any one time is due to the presence of strong local centres of change—many of these centres of change have a comparatively short life-time of the order of a hundred years.) It is also noticeable from Vestine's charts that the secular variation is much smaller in the Pacific hemisphere (120°E–80°W). This may be due to the scarcity of observations, which in any case do not go very far back in time. S. K. Runcorn (1956) has pointed out that if this feature is real it must reflect differences in the physical properties at depth within the Earth since it cannot be due to differences between an oceanic and continental crust. This would be a serious consequence since much of the physics of the Earth's interior is based on the model of a spherically symmetric Earth.

K. Whitham (1958) using Canadian isomagnetic and isoporic charts (epoch 1955.0) investigated, by a numerical variational method, the relationship between the observed secular variation and the non-dipole field in an area (Canada) where there are no isoporic foci. He found that the drift and decay of the non-dipole field corresponding to the subtraction of the axial dipole from the observed field cannot account for the observed secular variation in Canada—in particular there is no evidence for a westward drift of the expected magnitude. There is also no evidence of any northward drift. Whitham then re-examined that part of the data published by Bullard *et al.* (1950) appropriate for Canada and found evidence of a westward drift rate in recent years in Canada, but some three times smaller than the world-wide average. Thus significantly large changes in the rate of the westward drift can be found in a region occupying about 4 per cent of the surface of the Earth as compared to the world average. On the assumption that the westward drift reflects motions in the Earth's core, the decrease in westward drift in Canada would imply that the outer part of the core in this region is moving less slowly than the inside part of the core compared to the average motions throughout the core. This would indicate that in this region convective overturn is smaller. If the detailed regional nature of the westward drift could be determined, one might hope to be able to form some ideas about the structure of the convective motions in the core.

The westward drift of the geomagnetic field has been interpreted to imply that the outer core is rotating more slowly than the mantle. E. C. Bullard *et al.* (1950) related the westward drift to the effect of Coriolis forces on the motions in the core. They suggested that a differential angular velocity would be set up between the outer and inner core—the interchange of fluid across the outer core due to convection causing material with a smaller transverse velocity to rise to the top of the core. W. Munk and R. Revelle suggested later (1952) that the westward drift may be related to the irregular fluctuations in the length of the day. Irregularities in the rotation of the Earth fall into two classes—a wobble of the Earth and changes in the rate of rotation, i.e. changes in the length of the day. Figure 6.1 shows the spectrum of changes in the rate of rotation schematically arranged according to their time scale in years. There are a number of different peaks in the frequency spectrum—rapid changes of either sign being superposed on a substantially constant deceleration which is attributed to tidal friction in shallow seas. The time scale of the secular variation is extremely interesting in that it is intermediate between that of atmospheric and geologic events. It is

THE COMMONWEALTH AND INTERNATIONAL LIBRARY OF SCIENCE, TECHNOLOGY, ENGINEERING AND LIBERAL STUDIES

PERGAMON PRESS
OXFORD · LONDON · NEW YORK · PARIS

PERGAMON PRESS LTD.
Headington Hill Hall, Oxford
4 & 5 Fitzroy Square, London W.1

PERGAMON PRESS INC.
122 East 55th Street, New York 22, N.Y.

GAUTHIER-VILLARS
55 Quai des Grands-Augustins, Paris 6

PERGAMON PRESS G.m.b.H.
Kaiserstrasse 75, Frankfurt am Main

Set in Monotype Baskerville by Santype Ltd of Salisbury
and Printed in Great Britain by Tisbury Printing Works Ltd, Salisbury

Aims, Scope and Purpose of the Library

The Commonwealth and International Library of Science, Technology, Engineering and Liberal Studies is designed to provide readers, wherever the English language is used or can be used as a medium of instruction, with a series of low-priced, high-quality, soft-cover textbooks and monographs (each of approximately 128 pages). These will be up to date and written to the highest possible pedagogical and scientific standards, as well as being rapidly and attractively produced and disseminated—with the use of colour printing where appropriate—by employing the most modern printing, binding and mass-distribution techniques.

The books and other teaching aids to be issued by this Library will cover the needs of instructors and pupils in all types of schools and educational establishments (including industry) teaching students on a full and/or part-time basis from the elementary to the most advanced levels.

The books will be published in two styles—a soft-cover edition within the price range of 7*s*. 6*d*. to 17*s*. 6*d*. ($1.25 to $2.75) and a more expensive edition bound in a hard-cover for library use. The student, the teacher and the instructor will thus be able to acquire, at a moderate price, a personal library in whatever course of study he or she is following.

Books for Industrial Training Schemes to Increase Skills, Productivity and Earnings

To meet the ever-growing and urgent need of manufacturing and business organizations for more skilled workers, technicians, supervisors, and managers in the factory, the office and on the land, the Library will publish, with the help of trade associations, industrial training officers and technical colleges, specially commissioned books suitable for the various training schemes organized by or for industry, commerce and government departments. These books will help readers to increase their skill, efficiency, productivity and earnings.

New Concept in Educational Publishing; One Thousand Volumes to be Published by 1967: Speedy Translation and Simultaneous Publication of Suitable Books into Foreign Languages

The Library is a new conception in educational publishing. It will publish original books specially commissioned in a carefully planned series for each subject, giving continuity of study from the introductory stage to the final honours degree standard. Monographs for the post-graduate student and research workers will also be issued, as well as the occasional reprint of an outstanding book, in order to make it available at a low price to the largest possible number of people through our special marketing arrangements.

We shall employ the latest techniques in printing and mass distribution in order to acheive maximum dissemination, sales and income from the books published, including where suitable, their rapid translation and simultaneous publication in French, German, Spanish and Russian through our own or associated publishing houses.

The first 50 volumes of the Library will be issued by the end of 1962; during 1963 a further 150 titles will appear; we expect that by December 1967 a complete Library of over 1000 volumes will have been published. Such a carefully planned, large-scale project in aid of education is unique in the history of the book publishing industry.

New, Modern, Low-priced Textbooks for Students in Great Britain. International Co-operation in Textbook Writing, Publishing and Distribution

In those sciences such as Mathematics, Physics, Chemistry and Biology, which are started at an early age, there will be books suitable for students in Secondary, Grammar and Public Schools in Great Britain covering the work for the new Certificate of Secondary Education and the Ordinary and Advanced Levels of the General Certificate of Education. In all sciences there will be books to meet the examination requirements of the Ordinary National Certificate, Higher National Certificate, City and Guilds and the various other craft and vocational courses, as well as a full range of textbooks required for Diploma and Degree work at Colleges of Technology and Universities.

Similarly books and other teaching aids will also be provided to meet examination requirements in English-speaking countries overseas. Wherever appropriate the textbooks written for British students and courses will be made available to English-speaking students abroad and in particular in countries in the Commonwealth and in the United States.

The help of competent editorial consultants resident in each country will be available to authors at the earliest stages of the drafting of their books to advise them on how their volumes can be made suitable for students in different countries. If, because of differences in curricula and educational practice, substantial changes are needed to make a British textbook suitable in, for example, Australia, or a textbook written by an Indian or an American author suitable for use in the United Kingdom, then the Press will arrange for authors from both countries to collaborate to achieve this.

Co-publishers in the U.S.A. and Canada, The Commonwealth and Other Parts of the World, to Ensure Maximum Possible Dissemination at Low Prices

The Macmillan Company, New York, America's leading educational and general publishers, have already been appointed co-publishers of this Library in the U.S.A. and Canada, and negotiations are also in hand to appoint co-publishers in each of the major Commonwealth countries as well as in Europe, Africa and Asia, to market all books published in the Library exclusively in their country or territory. The co-publishers will assist authors and editors of the Library in the following ways:

(a) By making available to them their editorial contacts, resources and know-how to make the books commissioned for publication in the Library suitable for sale in their country.
(b) By purchasing a substantial quantity of copies of each book for exclusive distribution and sale.

(c) Where useful (in the interest of maximum dissemination), to arrange for or assist with the printing of a special edition, or the entire edition, of a particular textbook.

(d) To use their best endeavours to ensure that the books published in the Library are widely reviewed, publicized, distributed and sold at moderate prices throughout their marketing territory.

International Boards of Eminent Advisory, Consulting and Specialist Editors and Sponsoring Committee of Corporate Members

An Honorary Editorial Advisory Board and a Board of Consulting and Specialist Editors and a Sponsoring Committee under the chairmanship of Sir Robert Robinson, O.M., F.R.S., have been appointed. Some 500 eminent men and women drawn from all walks of life—Universities, Research Institutions, Colleges of Advanced Technology, Industry, Trade Associations, Government Departments, Technical Colleges, Public, Grammar and Secondary Schools, Libraries, and Trade Unions and parliamentarians interested in education—not only from this country but also from abroad—are available to advise by correspondence the editors, the authors, the Press and the national co-publishers to help achieve the high aims and purpose of the Library.

The launching of a library of this magnitude is a bold and exciting adventure. It comes at a time when the thirst for education in all parts of the world is greater than ever. Through education man can get an understanding of his environment and problems and a stimulation of interest which can enrich his life. And, too, if he learns how to apply the results of scientific research, material standards of life can be raised, even in a world of rapidly increasing population embroiled in a great arms race. Some of us hope and believe that through education lies the road to lasting world peace and happiness for all nations and communities, regardless of race, colour or ideology. In the history of education examples can be cited of how one or other famous textbook or author profoundly influenced the education of the period. When at some future time the history of education in the second half of the twentieth century is written, it may well be that the Commonwealth and International Library of Science, Technology, Engineering and Liberal Studies, published by Pergamon Press as a private venture with the co-operation of eminent scientists, educators, industrialists, parliamentarians and others interested in education, will stand out as one of the landmarks.

ROBERT MAXWELL
Publisher at Pergamon Press

Honorary Editorial Advisory Board

not unnatural therefore to consider a possible relationship with the remarkably large irregular fluctuations in the length of the day. Munk and Revelle have shown that these fluctuations cannot be due to any processes going on in the Earth's crust or mantle or

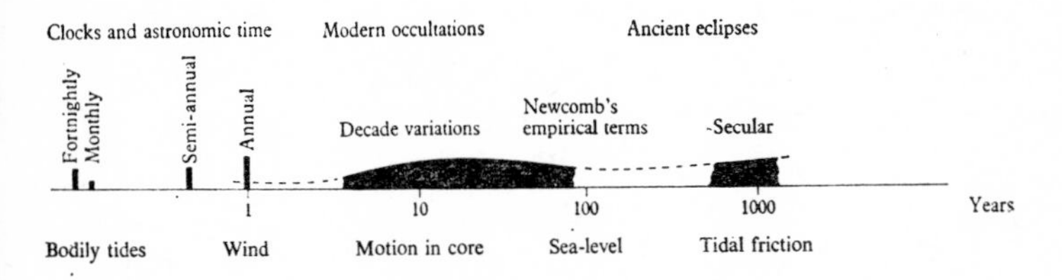

FIG. 6.1. Spectrum of changes in the rate of rotation of the Earth, schematically arranged according to their time scale in years. Vertical lines indicate discrete frequencies, shaded portions indicate a continuous spectrum. Principal source of the observations is shown above lines, and presumable geophysical cause beneath lines (after W. H. Munk and G. J. F. MacDonald).

in the oceans or atmosphere. No transport of mass at the Earth's surface which would alter the Earth's moment of inertia could account for such large changes. It is suggested that these irregular changes are due to electromagnetic torques adjusting the balance of angular momentum between the Earth's core and mantle.

Consider a model of the Earth in which the mantle and core rotate as two solid bodies (S. K. Runcorn 1954). The principal moments of inertia of the mantle and core are $I_M = 7{\cdot}2 \times 10^{44}$ g.cm^2 and $I_C = 0{\cdot}85 \times 10^{44}$ g.cm^2 respectively. If their angular velocities are Ω_M and Ω_C respectively, then conservation of angular momentum gives

$$I_M\Omega_M + I_C\Omega_C = \text{constant} \tag{6.1}$$

i.e.

$$I_M\delta\Omega_M + I_C\delta\Omega_C = 0 \tag{6.2}$$

The rate of change of westward drift is given by

$$-\delta\dot{\phi} = -(\delta\Omega_C - \delta\Omega_M) \tag{6.3}$$

From equations (6.2) and (6.3) it follows that

$$\delta\Omega_{\mathrm{M}} = -\frac{\delta\dot{\phi}}{1 + I_{\mathrm{M}}/I_{\mathrm{C}}} = -0{\cdot}105\delta\dot{\phi} \tag{6.4}$$

The reported increase in the rate of westward drift between the epochs 1905–1925 and 1925–1945 is $\delta\dot{\phi} = -0{\cdot}87 \times 10^{-10}$ rad/sec. Equation (6.4) then gives $\delta\Omega_{\mathrm{M}}/\Omega_{\mathrm{M}} = 12{\cdot}5 \times 10^{-8}$. On the other hand astronomical observations give $\delta\Omega_{\mathrm{M}}/\Omega_{\mathrm{M}} = 6 \times 10^{-8}$ for the period 1910–1930. Thus

$$\frac{(\delta\Omega_{\mathrm{M}})\text{ observed}}{(\delta\Omega_{\mathrm{M}})\text{ computed}} = 0{\cdot}48 \tag{6.5}$$

and the changes in the length of the day are thus not inconsistent with the reported fluctuations in the rate of westward drift—provided a substantial part of the outer core is involved in the fluctuations. The above calculation implies that there exists an adequate couple to produce the observed change in angular velocity. This could be caused by viscous or electromagnetic coupling. However, as Bullard (1950) has pointed out viscous coupling would lead to an eastward drift. Since the angular velocity of the mantle diminishes because of tidal friction, then, at any moment, the core will rotate slightly faster than the mantle and eastward with respect to it. This difficulty can be overcome by electromagnetic coupling. If there is an interchange of matter between the inner and outer parts of the core, the inner portion rotates somewhat faster and the outer somewhat slower than the core as a whole. The mantle is electromagnetically coupled to the entire core and the outer part moves westward with respect to it. The magnitude of the electromagnetic torque is difficult to estimate but is possibly sufficient—provided the most favourable values of the uncertain parameters are chosen.

It has been argued that the currents which produce the time-dependent part of the geomagnetic field must flow in a thin layer just below the boundary of the core. A layer of metal acts as a shield for variable currents and the depth of penetration d into a plane sheet of an electromagnetic wave of angular frequency Ω is given by

$$1/d = (2\pi\mu\sigma\Omega)^{\frac{1}{2}}$$

The situation in the Earth's core is opposite to that of the above skin effect, the field originating inside the conductor and penetrating from there to the outside. Putting $2\pi/\Omega = 100$ years, and $\sigma = 3 \times 10^{-6}$ e.m.u. leads to a value for d of 50 km. This is so small that it justifies the use of the plane-sheet approximation. However, as R. Hide and P. H. Roberts (1961) have emphasized this argument only holds if the core behaves as a rigid body. This is not so and they have shown that hydromagnetic waves can transmit magnetic changes from deep within the core to its surface both swiftly and with but little attenuation. The changing fields in the core induce currents in the mantle and their associated magnetic fields oppose the growth of the disturbance in the mantle and above. Much of the fine detail of the secular variation is thus most probably wiped out by the shielding action of the weakly conducting mantle.

There have been several suggestions that the moment of the equivalent dipole which most nearly fits the geomagnetic field began to increase some time after 1930. S. Procopiu (1947), C. Gaibar-Puertas (1951) and E. C. Bullard (1953) all suggested that a minimum occurred in the early 1930's. H. G. Macht reassessed the evidence and deduced a minimum in 1951–52, although T. Nagata and T. Rikitake (1957) expressed extreme doubt of the reality of such a minimum in the 1950's. This view has been confirmed by B. R. Leaton (1962) who obtained a value of $(-11 \pm 1)a^3\gamma$/year for the rate of change of the moment for the epoch 1955 where a is the mean radius of the Earth. It thus appears that the moment is still decreasing although perhaps at a slower rate, the average deduced from all spherical harmonic analyses from 1829 to 1955 being $-16{\cdot}5a^3\gamma$/year.

B. R. Leaton (1962) analysed British Admiralty Charts of the secular variation for the epoch 1955, and obtained the rate of westerly drift of the declination pattern given in Table 6.1. The marked dependence of the rate of drift on latitude is in substantial agreement with the results of the analysis by Bullard *et al.* (1950) of the non-dipole field. The relatively high rate of drift for 80°S is probably, but not entirely, due to difficulties in estimating secular changes at that latitude. The two extreme values of the variation of the westerly drift with longitude are also subject to uncertainties.

TABLE 6.1 *Annual Westerly Drift in Declination Pattern, Epoch 1955.0 (after B. R. Leaton)*

Lat. N.	Drift		Long. E.	Drift	
+80°	0·008°	±0·049°	0°	+0·155°	±0·020°
+70	0·045	±0·018	30	+0·147	±0·040
+60	0·052	±0·017	60	+0·250	±0·037
+50	0·077	±0·021	90	+0·143	±0·034
+40	0·090	±0·016	120	+0·156	±0·068
+30	0·134	±0·027	150	+0·131	±0·037
+20	0·170	±0·046	180	+0·169	±0·066
+10	0·210	±0·062	210	+0·045	±0·082
0	0·244	±0·103	240	−0·083	±0·061
−10	0·235	±0·118	270	+0·057	±0·032
−20	0·209	±0·098	300	+0·128	±0·020
−30	0·164	±0·138	330	+0·123	±0·019
−40	0·207	±0·112			
−50	0·164	±0·076			
−60	0·134	±0·048			
−70	0·115	±0·044			
−80	0·199	±0·031			

Leaton also obtained the annual rate of movement of the north magnetic dip pole for the epoch 1955.0 to be 0·059° ± 0·005° north and 0·084° ± 0·002° west in good agreement with the average annual movement during the first half of this century, 0·07° north and 0·09° west as found by K. Whitham and E. I. Loomer (1956). From a similar analysis from 9 points near the south magnetic dip pole Leaton obtained for its movements 0·027° ± 0·005° north and 0·175° ± 0·004° west.

6.4 Dynamo Models

There are two major problems to be solved in a dynamo theory of the Earth's main field. Firstly it is necessary to show that some pattern of hydrodynamical flow exists which can produce a (predominantly) axial dipole field. Secondly it must be shown that this flow exists in the Earth's core. There are two possible approaches to these problems. We may try and set up a model having a definite and suitably chosen set of motions. We then have to examine whether with the appropriate boundary conditions, the equations for a steady-state dynamo (see Appendix D and equation 6.6) possess solutions with a magnetic field external to the sphere in which the motions are occurring. The second method is not based on any particular model but considers the energy change between a magnetic field and a turbulent conducting fluid. W. M. Elsasser (1950) suggested that there is an equipartition between the average kinetic energy density and the average magnetic energy density. If there is equipartition, then taking a velocity of 0·03 cm/sec as observed for the secular variation near the surface of the core, we obtain a value for the magnetic field of 0·1 Γ which is too small by a factor of at least 100. It would follow then that either the fluid velocity increases rapidly with increasing depth in the core or else that the magnetic field is far above its equipartition value. On the other hand the remarkable stability of the dipole part of the geomagnetic field as compared to the higher harmonics favours a low mean velocity. Compared to the rapid changes of the higher harmonics, the inclination of the dipole axis relative to the Earth's rotational axis has not changed appreciably since the time of Gauss' first determination (1830). Also the longitude of the dipole axis has changed but slightly and seems to have remained nearly constant since about 1880. The very fact that a relatively large deviation of the dipole axis from the rotational axis can be maintained for over a century indicates that the lifetime of some major eddies must be fairly large—and this is hard to reconcile with velocities very much higher than those estimated at the surface of the core.

G. K. Batchelor (1950) has also shown that only in the smallest eddies might equipartition be expected to exist—although there is still some doubt on this point.

Sir Joseph Larmor (1919) originally suggested that through inductive interaction with a small inducing field parallel to the magnetic axis, steady meridional circulation of matter might produce zonal electric currents which amplify the inducing field. The effect of such a motion would be to stretch the lines of force within the core, the field outside remaining unchanged (see Fig. 6.2). T. G. Cowling (1934) showed, however, that steady motion

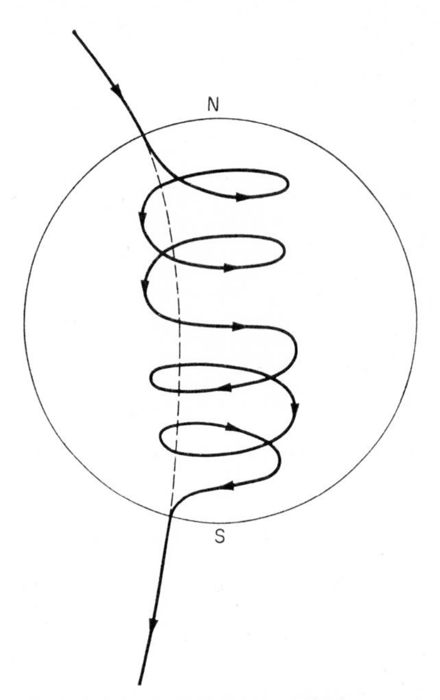

FIG. 6.2. (After R. Hide).

confined to meridian planes cannot amplify a field of the required type. W. M. Elsasser (1947) further showed that purely zonal motion parallel to latitude circles cannot amplify an inducing dipole field. Cowling's result has been further extended by G. E. Backus and S. Chandrasekhar (1956) and it appears that homogeneous dynamos must possess a low degree of symmetry. If convection alone was active, the fluid particles would describe closed curves confined to planes and no dynamo would result. If the system rotates, the paths of the particles are twisted into three-dimensional shapes by the action of the Coriolis force. Consider a spherical shell filled with fluid that is in radial convection (by heat developed on the inside, for example) and at the same time is rotating about an axis through its centre. If a magnetic field, whose lines of force are originally in meridional planes, exists within this non-uniformly rotating fluid, then the lines of force, being attached to the fluid particles, will be deformed in the manner shown in Fig. 6.3.

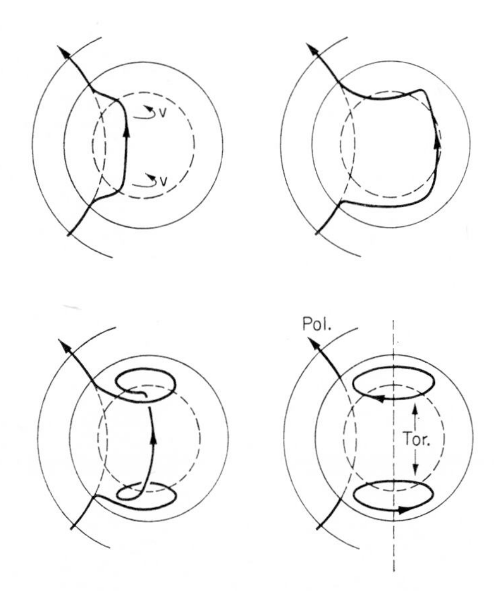

FIG. 6.3. Generation of a toroidal field in a non-uniformly rotating sphere (after W. M. Elsasser).

For a complete solution of the dynamo problem it would be necessary to solve equations (D.5) to (D.13) simultaneously with the correct boundary conditions. Needless to say no one has attempted this owing to the mathematical difficulties.* The method of attack has been to specify a velocity field **v** and then to try and solve equation (D.11) for the magnetic field **H**. In the case of a steady dynamo, equation (D.11) reduces to

$$\text{curl}\,(\mathbf{v} \times \mathbf{H}) + \nu_m \nabla^2 \mathbf{H} = 0 \tag{6.6}$$

Even when the velocity field **v** is assigned, the solution of (6.6) is still very difficult and only recently has it been proved that at least in principle steady dynamos can exist. (A. Herzenberg 1958; G. E. Backus 1958.)

Both E. C. Bullard and W. M. Elsasser have separated the electromagnetic and hydrodynamic problems and attempted to solve the former only, i.e. they assumed a particular motion in the Earth's core together with a magnetic field and calculated the electromagnetic interaction occurring within such a system. Elsasser (1946, 1947) used a series of spherical Bessel functions to represent the radial variation of the velocities and fields. E. C. Bullard and H. Gellman (1954) met with more success by assuming a perfectly general form for the radial functions, subject only to the equation of continuity (D.13).

A system of convection currents in a stationary sphere could involve motions of any of the S (poloidal) types. In a rotating sphere these would be accompanied by motions of the T (toroidal) type (see Appendix E). Figures 6.4 and 6.5 (after E. C. Bullard) show the general form of the fields for $n = 1, 2$. The S fields are more difficult to visualize than the T's since they have a radial component. A significant feature of the toroidal field is that it vanishes in an insulator or vacuum surrounding the conducting

* We do not know enough about the physical conditions prevailing deep within the Earth's interior, including the nature of the agency responsible for maintaining core motions, to be able even to formulate the boundary conditions under which these equations have to be solved.

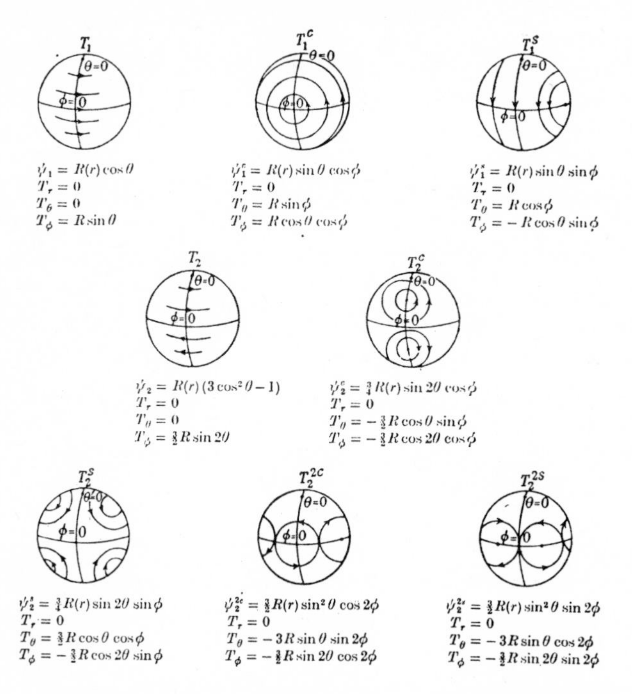

FIG. 6.4. Classification of toroidal fields (after E. C. Bullard).

sphere and so would not show up in measurements at the Earth's surface. The primary poloidal field decays and this implies that the secondary toroidal field must also ultimately decay no matter how strong it becomes during some finite time. A purely toroidal fluid motion does not interact with a toroidal field and a purely poloidal fluid motion merely rearranges the circular lines of force without generating a poloidal field. A general rotationally symmetrical motion is a linear combination of these two types—and hence for rotational symmetry no feedback process exists and there can be no dynamo of full rotational symmetry. Elsasser and Bullard have drawn up tables showing what interactions are possible between different types of fluid motions and a given field.

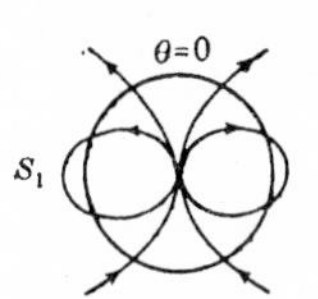
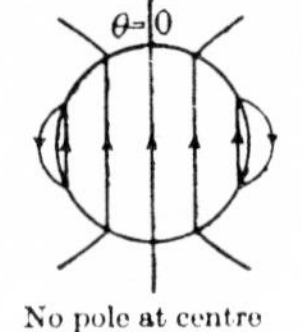
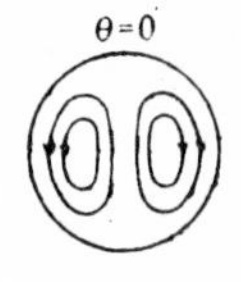

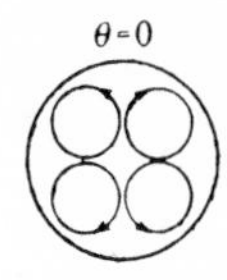

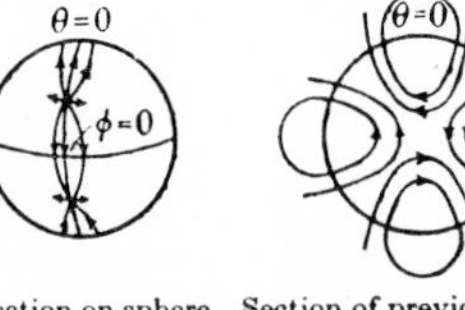
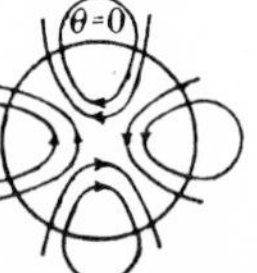
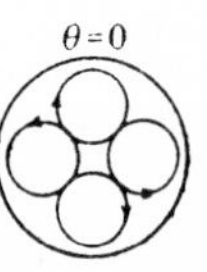

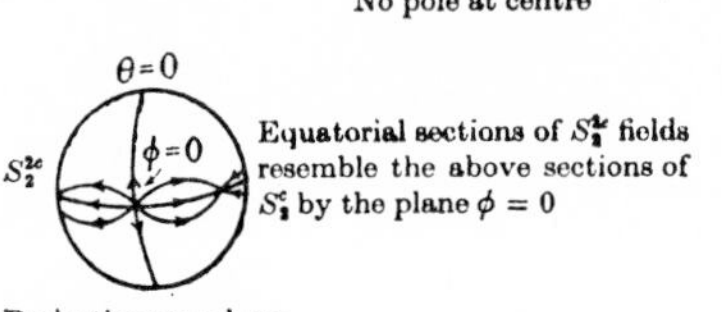

FIG. 6.5. Classification of poloidal fields (after E. C. Bullard).

It appears that in order to produce the poloidal field outside the Earth a much more powerful toroidal field must exist in the Earth's core. Since the field outside the Earth is predominantly S_1 it is natural to consider what interactions could stem from it. It is found that neither the combination of motions T_1S_1 nor T_1S_2 produce a closed chain returning to S_1 and so cannot maintain a field. On the other hand, the combinations $T_1S_2^{2c}$, $T_1S_2^{c}$ and $T_1S_1^{c}$ (and the corresponding combinations with s in place of c in the upper index) do produce closed chains. The combination

$T_1S_2^{2c}$ was chosen for a detailed study. E. C. Bullard and H. Gellman replaced the differential equations by finite difference approximations, resulting in a set of linear homogeneous algebraic equations which they solved with the aid of electronic computers. After a considerable amount of computation, they were able to show that the system could act as a dynamo, although it must be borne in mind that this result does not in itself imply that the correct solution has been obtained. H. Takeuchi and Y. Shimazu (1953) have also examined the problem from this view point and concluded that the self-exciting dynamo is possible by a variety of fluid motions, and that the maximum radial velocity of the field is of the order 10^{-2} cm/sec.

When energy is transferred from a moving fluid to a magnetic field, we can speak of "amplification". The simplest example is illustrated in Fig. 6.6. Assume that originally there is a magnetic

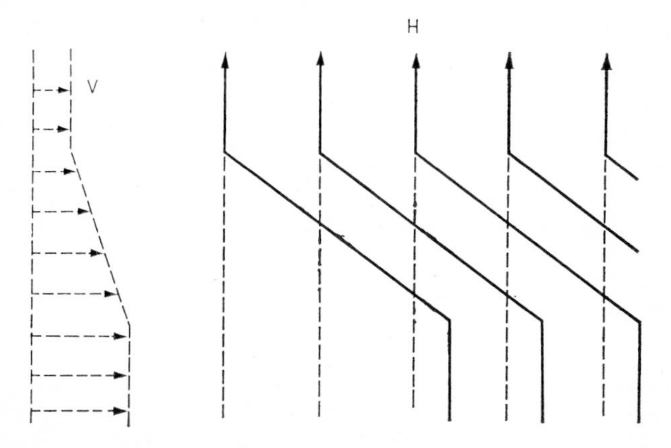

FIG. 6.6. Amplification of a magnetic field by a linear velocity shear normal to the field (after W. M. Elsasser).

field in the y direction (dashed lines) and that at time $t = 0$ a fluid motion is started that is in the x direction but which has a gradient in the y direction as indicated in the velocity profile on the left-hand side of the figure. In the absence of free decay the lines of force are deformed into the shape shown by the heavy lines. An x-component of the magnetic field is generated whose magnitude increases with time. The total energy content of the magnetic field has increased at the expense of the work done by the kinetic energy of the fluid. Eventually the fluid motion will be slowed down by the pondero-motive forces exerted by this field. The process indicated above is obviously capable of many variations depending on the geometry of the fluid motion and original magnetic field.

The existence of the westerly drift of the secular variation indicates that the core in all probability does not rotate uniformly as a rigid body but that its angular velocity varies with depth. Such a motion will deform the magnetic lines of force of the dipole field as shown schematically in Fig. 6.3. Assume that we

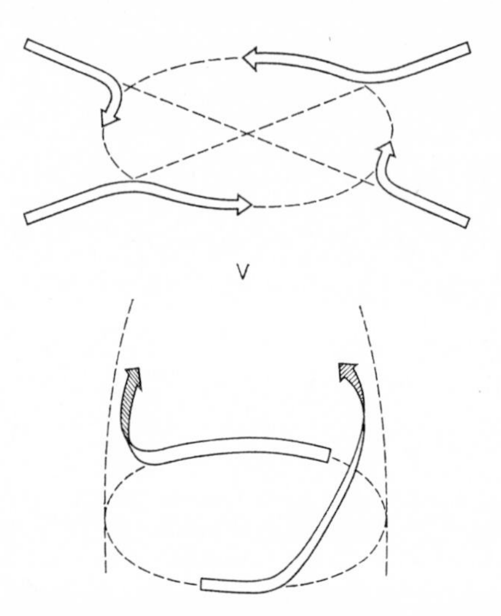

FIG. 6.7. Coriolis effect on a locally converging and rising eddy of fluid (after E. N. Parker).

first have a dipole line of force in a meridional plane (dashed line) and that the inner part of the core rotates relative to the outer.

The lines of force then get dragged around circles of latitude as shown in Fig. 6.3 and if this non-uniform rotation continues, the lines of force will become wrapped around the axis of rotation in approximately a system of circles. The most significant feature of this particular mechanism is its lack of reciprocity—there is no similar interaction whereby starting from a toroidal field one can produce a poloidal field of the type represented by the original dipole. (The lines of force of a toroidal field are circles centred on the axis. Any rotationally symmetrical motion transforms a family of circles into another such family and this transforms any toroidal field into another toroidal field.) We thus have no dynamo model—we need a process which maintains the primary poloidal field. As already pointed out we must consider fluid motions that are not symmetrical about the axis of rotation—a hydromagnetic dynamo must therefore be essentially three-dimensional.

E. N. Parker (1955) considered the effect of the Coriolis force on local convective eddies. Suppose we have a series of streams of fluid rising radially outward and sinking elsewhere, and consider in particular one such rising stream in the direction of the Earth's rotational axis. At its lower end there must be lateral convergence and at its upper end divergence of the fluid. As the fluid converges it will be deflected (to the right in the northern hemisphere) and at the same time it rises so that the net result will be a spiralling motion (see Fig. 6.7). At the top of the stream the fluid diverges and the Coriolis force acting now in the opposite sense will uncoil the spiral again. If the convective stream is not along the Earth's rotational axis, the geometry of the spiralling motion will, of course, be more complicated.

Consider now the deformation of a toroidal magnetic field by such a spiral (Fig. 6.8)—the axis of the convective stream being normal to the field lines. The field lines become lifted and at the same time undergo a circular twist. If the twist is 90°, a closed loop of magnetic force is created in a plane perpendicular to the original lines. Parker has shown how these loops may coalesce to form an overall poloidal field (see Fig. 6.9). To obtain the

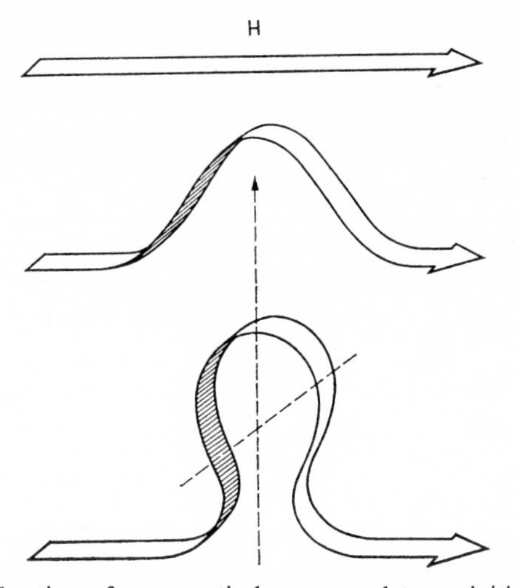

FIG. 6.8. Creation of a magnetic loop normal to an initial toroidal field by means of the local motion shown in Fig. 6.7 (after E. N. Parker).

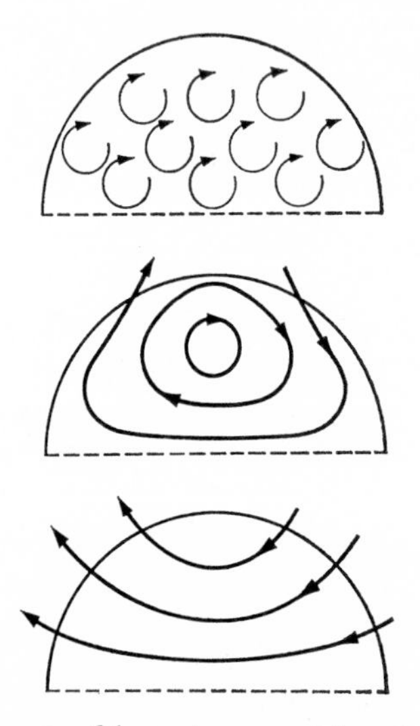

FIG. 6.9. Coalescence of loops to regenerate a poloidal dipole field.

correct sign for amplification of the poloidal field it is necessary that the convective motions are asymmetrical with respect to rising and sinking eddies. Moreover the formation of loops suitable for feedback occurs only when the angle of rotation of the eddy is not too different from 90°. This is due to the mechanical forces that counteract eddy formation, the most important of which is the pondero-motive force set up by the deformation of the toroidal magnetic field. It must also be pointed out that such a dynamo is not a stationary generator—they are only stationary in the mean, since individual eddies must appear and then die out after having twisted the toroidal field lines by a suitable amount.

In 1958 G. Backus and A. Herzenberg, working independently, each showed that it was possible to postulate a pattern of motions in a sphere filled with a conducting fluid in such a way that the arrangement acts as a dynamo producing a magnetic field outside the conductor. In each case the motions were physically very improbable; however, rigorous mathematical solutions were obtained, as was not the case with Bullard's numerical solution. The motions obtained by Backus all involved periods when the fluid is at rest. He needs these periods of rest to ensure that other fields generated by induction will not develop in such a fashion that they eventually destroy the whole process. Backus considered an azimuthal shearing motion symmetric about the axis of rotation which proceeds long enough to produce from any arbitrarily weak poloidal field a very large toroidal field. The motion is now supposed to cease and ohmic losses lead to a decay of all fields present, the higher harmonics decaying more rapidly. After a sufficient period of time the only significant field will be the toroidal decay field where almost any velocity which has a radial component and is not axisymmetric will regenerate the initial poloidal field and external dipole moment. The model of the core obtained by Herzenberg consists of two spheres (which may be pictured as two eddies) each of which rotates as a rigid body at a constant angular velocity about a fixed axis. About a half of all possible configurations can act as dynamos if the velocities, positions, and radii of the rotating spheres are suitably adjusted. The essential point of Herzenberg's dynamo is that the axially symmetric component of the magnetic field of one of the spheres is twisted by the rotation resulting in a toroidal field which is strong enough to give rise to a magnetic field in the other sphere. The axial component of this field is twisted as well and fed to the first sphere. If the rotation of the spheres is sufficiently rapid, a steady state may be reached.

Since the geomagnetic field exhibits a secular variation, the dynamo action is certainly not steady. Palaeomagnetic evidence indicates that the dipole field may even reverse its polarity (see Chapter 7). The mathematical difficulties of non-linear differential equations have precluded any general theory of non-steady dynamos.

Although normal and reversed fields could equally well exist according to the dynamo theory of the Earth's main field, it has not been demonstrated until recently (D. W. Allan, 1958) whether a means of transfer from one state to another could exist. E. C. Bullard (1955) examined the time-dependent behaviour of a disc dynamo—he replaced the fluid motion by a rotating, conducting disc, the pondero-motive force being obtained from the electromagnetic coupling between the disc and an electric coil connecting the rotating axis to the edge of the disc through brushes. Bullard showed that oscillations could be maintained even in the presence of ohmic dissipation and small viscous forces. The magnetic field, however, never reverses once it has been established in a given direction. T. Rikitake (1958) extended Bullard's analysis to a series of disc dynamos, each coupling with that at the next stage, the magnetic field produced by the last stage being fed back to the first. For the case of two disc dynamos, Rikitake showed that reversals of electric current and hence magnetic field can occur rapidly and at irregularly spread intervals. D. W. Allan (1958) extended the numerical integration of Rikitake's two coupled disc dynamos and showed that reversals can occur under a wide range of conditions. He showed that the phenomenon of reversal can occur, both in the sense of a change of polarity and in the sense of a true reversal of behaviour corresponding to a change from oscillations about one state of equilibrium to those about another. T. Yukutake (1960) considered the non-steady state of Herzenberg's dynamo for slowly changing magnetic fields. He found (also with the aid of an electronic computer) that Herzenberg's dynamo is unstable for small disturbances, and calculated the characteristic periods of changes for likely size eddies in the Earth's core. In a long series of papers, T. Rikitake (1955, 1956, 1958, 1959) showed that Bullard's dynamo is unstable for small disturbances if the Earth's rotation is neglected, although if rotation is taken into account, it appears stable.

References

ALLAN, D. W. Reversals of the Earth's magnetic field. *Nature* **181**, 469 (1958).

BACKUS, G. E. A class of self sustaining dissipative spherical dynamos. *Ann. Phys.* **4**, 372–447 (1958).

BACKUS, G. E. and CHANDRASEKHAR, S. On Cowling's theorem on the impossibility of self-maintained axi-symmetric homogeneous dynamos. *Proc. Nat. Acad. Sci.* **42**, 105–9 (1956).

BATCHELOR, G. K. On the spontaneous magnetic field in a conducting liquid in turbulent motion. *Proc. Roy. Soc. A.* **201**, 405–17 (1950).

BATE, G. L., POTRATZ, H. A. and HUIZENGA, J. R. Thorium in iron meteorites. *Geochim. et Cosmochim. Acta* **14**, 118–25 (1958).

BULLARD, E. C. The magnetic field within the Earth. *Proc. Roy. Soc. A.* **197**, 433–53 (1949).

BULLARD, E. C. Electromagnetic induction in a rotating sphere. *Proc. Roy. Soc. A.* **199**, 413–43 (1949).

BULLARD, E. C. The transfer of heat from the core of the Earth. *Mon. Not. Roy. Astr. Soc. Geophys.* Suppl. 6, 36–41 (1950).

BULLARD, E. C. Is the Earth's dipole moment increasing? *J. Geophys. Res.* **58**, 277–8 (1953).

BULLARD, E. C. The stability of a homopolar dynamo. *Proc. Camb. Phil. Soc.* **51**, 744–60 (1955).

BULLARD, E. C., FREEDMAN, C., GELLMAN, H. and NIXON, J. The westward drift of the Earth's magnetic field. *Phil. Trans. Roy. Soc. A.* **243**, 67–159 (1950).

BULLARD, E. C. and GELLMAN, H. Homogeneous dynamos and terrestrial magnetism. *Phil. Trans. Roy. Soc. A.* **247**, 213–78 (1954).

COWLING, T. G. The magnetic field of sunspots. *Mon. Not. Roy. Astr. Soc.* **94**, 39–48 (1934).

Elsasser, W. M. On the origin of the Earth's magnetic field. *Phys. Rev.* **55**, 489–98 (1939).

Elsasser, W. M. Induction effects in terrestrial magnetism. Part I: Theory. *Phys. Rev.* **69**, 106–16 (1946).

Elsasser, W. M. Induction effects in terrestrial magnetism. Part II: The secular variation. *Phys. Rev.* **70**, 202–12 (1946).

Elsasser, W. M. Induction effects in terrestrial magnetism. Part III: Electric modes. *Phys. Rev.* **72**, 821–33 (1947).

Elsasser, W. M. Causes of motions in the Earth's core. *Trans. Amer. Geophys. Union* **31**, 454–62 (1950).

Elsasser, W. M. The Earth's interior and geomagnetism. *Rev. Mod. Phys.* **22**, 1–35 (1950).

Elsasser, W. M. Hydromagnetism I: A review. *Amer. J. Phys.* **23**, 590–609 605, Fig, 4 (1955)

Gaibar-Puertas, C. Varacion secular del campo geomagnetico. *Observ. del Ebro*, Memo. No. 11, 1953.

Herzenberg, A. Geomagnetic dynamos. *Phil. Trans. Roy. Soc. A.* **250**, 543–85 (1958).

Jacobs, J. A. Temperature–pressure hypothesis and the Earth's interior. *Can. J. Phys.* **31**, 370–6 (1953).

Larmor, J. How could a rotating body such as the sun become a magnet? *Rep. Brit. Ass.* 159–160 (1919).

Leaton, B. R. Geomagnetic secular variation for the epoch 1955.0. *Roy. Obs. Bull.* No. 57, London (1962).

Lowes, F. J. Secular variation and the non-dipole field. *Ann. Geophys.* **11**, 91–94 (1955).

Macht, H. G. On the increase of the Earth's dipole moment. *J. Geophys. Res.* **59**, 369–76 (1954).

Munk, W. and Revelle, R. On the geophysical interpretation of irregularities in the rotation of the Earth. *Mon. Not. Roy. Astr. Soc. Geophys. Suppl.* 6, 331–47 (1952).

Munk, W. H. and MacDonald, G. J. F. *The rotation of the Earth.* Cambridge Univ. Press (1960).

Nagata, T. and Rikitake, T. Geomagnetic secular variation during the period from 1950 to 1955. *J. Geomag. Geoele.* **9**, 42–50 (1957).

Parker, E. N. Hydromagnetic dynamo models. *Astrophys. J.* **122**, 293–314 (1955).

Reed, G. W. and Turkevich, A. The uranium content of two iron meteorites. *Natl. Acad. Sci.–Natl. Res. Coun. Publ.* 400, 97–99 (1956).

Rikitake, T. Magneto hydrodynamic oscillations in the Earth's core. *Bull. Earthquake Res. Inst.* **33**, 1–25 (1955).

Rikitake, T. Magneto hydrodynamic oscillations of a conducting fluid sphere in a uniform magnetic field. *Bull. Earthquake Res. Inst.* **33**, 175–98 (1955).

Rikitake, T. Growth of the magnetic field of the self-exciting dynamo in the Earth's core. *Bull. Earthquake Res. Inst.* **33**, 571–82 (1955).

Rikitake, T. Magneto-hydrodynamic oscillations of finite amplitude of a conducting fluid sphere. *Bull. Earthquake Res. Inst.* **33**, 583–92 (1955).

RIKITAKE, T. Magneto-hydrodynamic oscillations of a conducting fluid sphere under the influence of the Coriolis force. *Bull. Earthquake Res. Inst.* **34**, 139–56 (1956).

RIKITAKE, T. Stability of the Earth's dynamo. *Bull. Earthquake Res. Inst.* **34**, 283–9 (1956).

RIKITAKE, T. Oscillations of a system of disc dynamos. *Proc. Camb. Phil. Soc.* **54**, 89–106 (1958).

RIKITAKE, T. Forced oscillations of the Earth's dynamo. *Bull. Earthquake Res. Inst.* **37**, 245–64 (1959).

RIKITAKE, T. Thermo-magneto-hydrodynamic oscillations in the Earth's core. *Bull. Earthquake Res. Inst.* **37**, 405–22 (1959).

RUNCORN, S. K. The Earth's core. *Trans. Amer. Geophys. Union* **35**, 49–63 (1954).

RUNCORN, S. K. The magnetism of the Earth's body. *Handbuch der Physik* Band XLVII, 498–533, Springer-Verlag (1956).

TAKEUCHI, H. and SHIMAZU, Y. On a self-exciting process in magneto-hydrodynamics. *J. Geophys. Res.* **58**, 497–518 (1953).

UREY, H. C. *The Planets.* Yale Univ. Press (1952).

VESTINE, E. H., LAPORTE, L., COOPER, C., LANGE, I. and HENDRIX, W. C. Description of the Earth's magnetic field and its secular change. *Carnegie Inst. Wash. Publ.* No. 578 (1947).

VESTINE, E. H., LAPORTE, L., LANGE, I. and SCOTT, W. E. The geomagnetic field: Its description and analysis. *Carnegie Inst. Wash. Publ.* No. 580 (1947).

WHITHAM, K. The relationships between the secular change and the non-dipole fields. *Can. J. Phys.* **36**, 1372–96 (1958).

WHITHAM, K. and LOOMER, E. I. The diurnal and annual motions of the north magnetic dip pole. *J. Atmos. Terr. Phys.* **8**, 349–51 (1956).

YUKUTAKE, T. Stability and non-steady state of self-exciting dynamos. Part I: *Bull. Earthquake Res. Inst.* **38**, 1–12 (1960). Part II: *Bull. Earthquake Res. Inst.* **38**, 437–49 (1960).

Additional Reading

HIDE, R. The hydrodynamics of the Earth's core. *Physics and Chemistry of the Earth.* Vol. 1, pp. 94–137. Pergamon Press (1956).

HIDE, R. and ROBERTS, P. H. The origin of the main geomagnetic field. *Physics and Chemistry of the Earth*, Vol. 4, pp. 27–98. Pergamon Press (1961).

CHAPTER 7

Palaeomagnetism

7.1 Introduction

One of the main difficulties in any geophysical investigation is that observations of natural phenomena cover only a very small fraction of the Earth's lifetime—less than about one part in 10^6. With such an extremely small sample, it is not to be wondered at that our knowledge of many geophysical processes is so rudimentary. Two fields of research however have yielded information about the early history of the Earth. These are isotopic studies and palaeomagnetism. The radioactive decay of certain rocks and minerals has systematized geochronological studies—particularly the dating of older rocks and the age of the Earth itself. A study of the variation of non-radioactive isotopes has also proved extremely valuable. One example which was pioneered by H. C. Urey in 1951 is the determination of palaeotemperatures by measuring the ratio of the oxygen isotopes O^{18}/O^{16}. There have also been enormous strides in the development of palaeomagnetic studies especially during the past decade. This is far too large a subject to be discussed in detail in this book, although certain aspects will be considered in this chapter—in particular that of reversals of polarity of the Earth's magnetic field.

An example of the way in which magnetic studies can give exciting new information on the tectonic history of the Earth is shown in the results of a magnetic survey of part of the Pacific Ocean. Scripps Institution of Oceanography conducted such a survey during 1955–1956 off the west coast of the United States of America between latitudes 32° and 36°N and longitudes 121° and 128°W. Prior to this, they had towed a magnetometer on several

Pacific cruises. In all of them strong magnetic anomalies were found that could be correlated with topography except over certain ridges and widely scattered sea mounts. The survey was caried out by a magnetometer towed behind the U.S. Coast and Geodetic survey ship *Pioneer* with as much detail and accuracy as an airborne magnetometer survey over land. The area was covered with a regular grid of accurately surveyed lines, usually only 5 miles apart. The probable absolute error of a single observation was about 10γ, although the relative error between any two points on the same line is negligible. The resulting total magnetic intensity map (Fig. 7.1) shows a series of narrow anomalies of about 400γ amplitude trending N–S over a substantially flat abyssal plain. No dry land survey has ever revealed a lineation that aproached this in uniformity and extent. The anomalies arise at shallow depths beneath the ocean bottom and are consistent with the presence of relatively thin layers of magnetized material overlying a less magnetic crust; they could equally well arise from topography of the upper surface of a continuous magnetic basement for the most part concealed by an overlying blanket of sediments. There can be little doubt, however, that the major positive anomalies are underlain by an excess, probably totalling at least 1 km in thickness, of a material highly magnetic by comparison with adjacent formations and almost certainly a basic igneous rock. The most obvious and acceptable geological explanation is that the excess material represents lava flows which have spread out over the floor of the ocean possibly filling pre-existing troughs. Within the limits of the survey the major anomalies cover an area of about 50,000 km^2. In extent therefore, and in thickness, the flows would rank with the largest continental lava fields, such as the Columbia River basalts which cover an area of 120,000 km^2 and in places reach thicknesses approaching 2 km. One of the most interesting features of the magnetic map is that the anomalies are intersected by the Murray fracture zone which interrupts their pattern. Whatever their origin and form, the rock bodies which give rise to the N–S

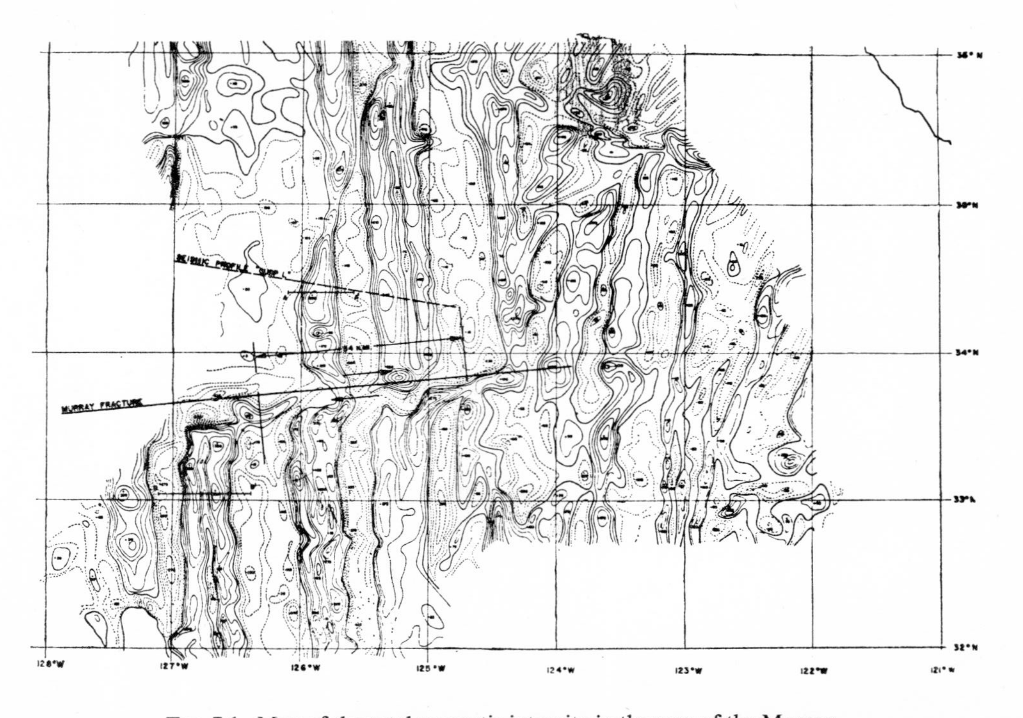

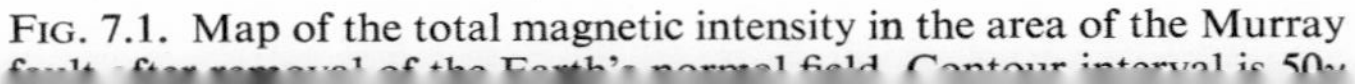

FIG. 7.1. Map of the total magnetic intensity in the area of the Murray

anomalies indicate lines of weakness in the ocean floor which were active prior to the formation of the Murray fracture zone. An offset of the pattern of the magnetic features as they cross it suggests a right lateral displacement of 84 nautical miles along the Murray fracture. This displacement, the N–S lineation of the anomalies and the right lateral displacement along the San Andreas fault, may well be related by a single structural development, possibly connected with a westerly movement of the North American continent relative to the floor of the Pacific.

Two other large breaks in the N–S magnetic lineation were found, across the Pioneer fault and the Mendocino fault. From the records of this first survey, it was not possible to match the contour lines of these faults however much they were shifted. In 1958 additional magnetic readings were made farther west on a few tracks north and south of the Pioneer fault. Finally a match was found revealing a displacement of more than 130 nautical miles with the north side of the fault shifted to the west with respect to the south side. The next year, the Scripps ship *Baird* extended the magnetic survey several hundred miles to the west, north and south of the Mendocino fault but without locating a match. The following spring corresponding sections of the pattern were finally found showing that the north side had shifted to the west with respect to the south side by approximately 600 miles. Figure 7.2 shows magnetic profiles across the Mendocino and Pioneer faults, the vertical solid lines being drawn through matching sections of the records. The top block contains profiles taken north of the Mendocino fault; the middle block, profiles south of the Mendocino and north of the Pioneer fault; and the bottom block, profiles south of the Pioneer fault. The dashed profile in the bottom block is the topmost profile inserted, after a shift of more than 700 miles, for comparison. These magnetic surveys thus give convincing proof of large-scale movements of parts of the Earth's crust, and the possibility of such large displacements must be considered in all tectonic studies.

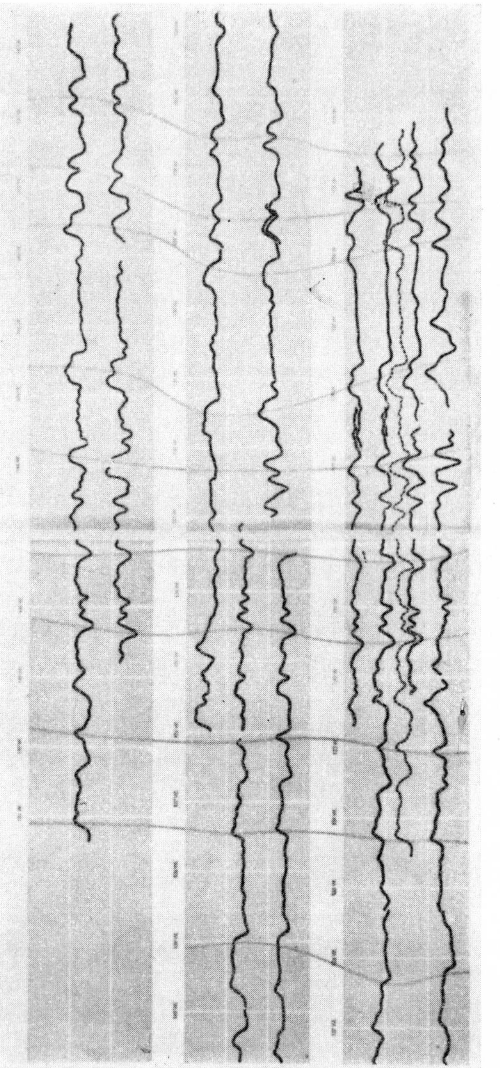

FIG. 7.2. Matching magnetic profiles across the Mendocino and Pioneer faults (see page 95) (after A. D. Raff).

7.2 The Magnetization of Rocks

The classic early work in palaeomagnetism is that of R. Chevallier who showed that the remanent magnetizations of several lava flows on Mount Etna were parallel to the Earth's magnetic field measured at nearby observatories at the time the flows erupted. Since then much work has been done in an effort to understand the processes by which rocks become magnetized. Most rocks owe their magnetism to the various oxides of iron which they contain. However, the manner in which igneous and sedimentary rocks acquire their permanent magnetism is very different.

Igneous rocks have been injected into pre-existing rock and extruded at the surface in a liquid state and then subsequently cooled and solidified. When lava cools and freezes following a volcanic out-burst, it will acquire a permanent magnetization dependent on the orientation and strength of the geomagnetic field at that time. This magnetization may remain practically constant because of the small capacity for magnetization in the Earth's field after freezing, and is much larger than would be acquired in the present geomagnetic field at 20°C. It has been established experimentally that the permanent magnetization acquired by an igneous rock which has been heated and cooled through the Curie point in magnetic fields of the order of the Earth's field (this permanent magnetization is called thermo-remanent magnetization—TRM) is considerably greater than the permanent magnetization acquired by exposing it to the same field at room temperature.

Sedimentary rocks may be magnetized by the orientation in the Earth's field of small grains of magnetic material such as magnetite when deposited in shallow rivers, lakes and seas. The magnetic grains tend to align themselves along the direction of the Earth's field while the sediment is still wet and unconsolidated. Later the material may harden because of compression and the magnetic grains may get locked in their original positions. In addition many sedimentary rocks undergo marked physical and chemical changes while they are being consolidated by compression and perhaps subjected to heating. One of the principal uncertainties in the interpretation of palaeomagnetic data is that of deciding how much the rocks have changed physically and chemically since they were laid down.

Any theory of the Earth's magnetic field must depend on whether present measurements are representative of the characteristics of the field in the past. We would like to know the answers to such questions as to whether the average field intensity has varied significantly over geologic time, whether the field has always been approximately that of an axial dipole and if so whether the polarity has always been the same, and whether the present characteristic features of the secular variation are permanent or transient.

Compared to the large number of ancient field directions which go back to Pre-Cambrian times, less than 10 accurate determinations have been made of the intensity of the geomagnetic field in the past—and the oldest only goes back to about 600 B.C. The few valid intensity measurements are mainly the result of the researches of E. and O. Thellier. In their review article (1959), they describe the difficulties of the problem and the reasons for such a small number of intensity measurements. Their results indicate a decrease in the main dipole moment of the geomagnetic field since well before the period of direct measurements. They obtained an average rate for the decrease which was about one half that indicated from spherical harmonic analyses of the field over the past century. Recent work by S. P. Burlatskaya (1962) indicates that the absolute magnitude of the Earth's field may be cyclic, the duration of a cycle being of the order of 10,000 years. His results suggest that the magnitude increased until about the beginning of our era, after which it began to decrease. With regard to the question of whether the Earth's field has always been approximately that of an axial dipole, striking statistical evidence that it was axisymmetrical in Mesozoic times (5–50 million years ago) has been obtained by O. W. Torreson *et al.* (1949). Figure 7.3

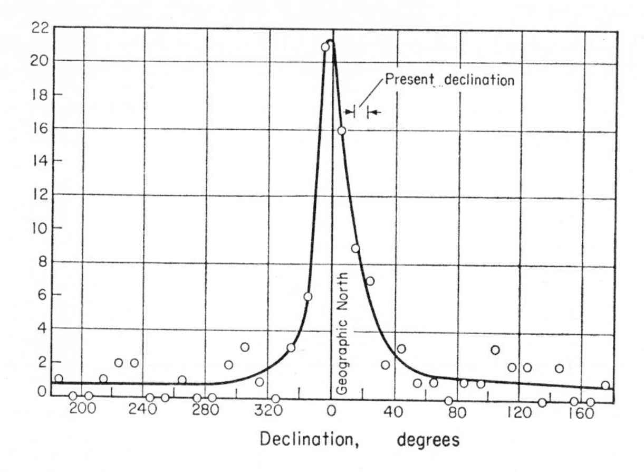

FIG. 7.3. Frequency distribution of declination measurements on Mesozoic rock samples (after O. W. Torreson, T. Murphy and J. W. Graham).

shows how the mean declination of flat lying sedimentary rocks of Mesozoic age gives a better fit to an axial dipole than to the present dipole. The dipolar character of the Earth's field in the past has been indicated by the grouping of pole positions for Tertiary rocks around the present geographical pole (see, for example, the work of J. Hospers (1951) on Icelandic rocks). The question of the polarity of the dipole will be discussed in Section 7.3.

If it is assumed that the geomagnetic field at the Earth's surface averaged over several thousands of years can be represented by a geocentric dipole with its axis along the axis of rotation, it is possible, by measuring the present direction of magnetization of a suite of rocks, to deduce the position of the Earth's rotational axis relative to the location of the rocks at the time when they were laid down. The measured declination will give the azimuth of the land mass at the time the rock became magnetized and the measured inclination will give the geographical latitude (see equation 5.6). It is thus possible to calculate ancient pole positions.

The accuracy of the results will depend on the accuracy with which the rocks can be dated and on the reliability of the magnetic measurements which in turn depend on the stability of the magnetization. An interpretation of such palaeomagnetic results involves the hypotheses of polar wandering and continental drift. These intriguing and controversial questions unfortunately lie outside the scope of this book.

7.3 Reversals of the Earth's Magnetic Field

Many rock samples of different ages and from different localities have shown a magnetic polarization opposite to that of the present field. Reverse magnetization was first discovered in 1906 by B. Brunhes in a lava from the Massif Central mountain range in France. Since then examples have been found in almost every part of the world including Spitsbergen, Greenland, Europe, Australia, Japan and South Africa where H. Gelletich (1934–1936) discovered that the huge Pilansberg dyke system was reversely magnetized. After the last war the problem was investigated in some detail by J. M. Bruckshaw and E. I. Robertson (1949) on dykes in northern England, by A. Roche (1950, 1951, 1953) on Tertiary lava flows in central France, by J. Hospers (1951, 1953, 1954) on Tertiary lava flows in Iceland and by C. D. Campbell and S. K. Runcorn (1956) on the late-Tertiary lavas of the Columbia River basalts. In each case it was found that about one-half of the flows were normally and one-half reversely magnetized. The results of Hospers' work are given in Table 7.1. Roche in his studies in France listed 15 cases of reversely magnetized flows and 11 of normal magnetization. Neither Roche nor Hospers could detect any differences by laboratory experiments between the magnetic properties of the normal and reversed rocks, and both concluded that the Earth's field reversed its polarity several times during the Tertiary—probably about every half-million years—the duration of the reversal process being of the order of 10,000 years.

There is no *a priori* reason why the Earth's field should have a particular polarity, and in the light of modern views on its origin there is no fundamental reason why its polarity should not change. Periodic field reversals, with the last reversal ending in early Quaternary time, were actually first proposed by M. Matuyama in 1929. There have been many cases where reversely magnetized

TABLE 7.1. *Magnetization of Lava Flows in Iceland**

Approx. age, years	Polarization
0–200	Normal
2000–7000	Normal
150,000	Normal
500,000	Normal
(0·5–1) × 10^6	Reverse
(2–12) × 10^6	Normal Reverse
(12–26) × 10^6	Reverse Normal Reverse Normal

* After J. Hospers.

lava flows cross sedimentary layers. Where the sediments have been baked by the heat of the cooling lava flow, they were also found to be strongly magnetized in the same reverse direction as the flow (see Fig. 7.4). In fact in all reported cases the direction of magnetization of the baked sediment is the same as that of the dyke or lava which heated it, whether normal or reversed. It seems improbable that the adjacent rocks as well as the lavas themselves would possess a self-reversal property and such results seem difficult to interpret in any other way than by a reversal of the Earth's field. However, before such an explanation is accepted, it must be asked whether there exist any physical or chemical processes whereby a material could acquire a magnetization opposite in direction to the ambient field. J. W. Graham (1949) found some sedimentary rocks of Silurian age which were reversely magnetized. He was able to identify the precise geological

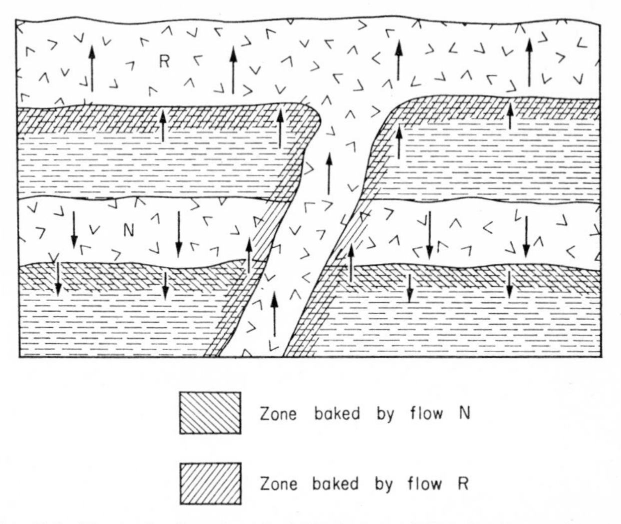

FIG. 7.4. Magnetic directions in baked zones (after A. Cox and R. R. Doell).

horizon over a distance of several hundred miles by the presence of a rare fossil which only existed during a short geological period. He found that some parts of the horizon were normally magnetized and some reversely and argued that this could not be accounted for by a reversal of the Earth's field which would affect all contemporaneous strata alike. (However, if the time scale of reversals in Silurian times was as short as Hospers found for Tertiary rocks, then Graham's fossil might well have survived at least one reversal.) Graham thus wrote to Professor L. Néel of Grenoble and asked him if he could think of any process by which a rock could become magnetized in a direction opposite to that of the ambient field. Néel came up with not one but four possible mechanisms—and within two years two of these four mechanisms had been verified, one by T. Nagata for a dacite pumice from Haruna in Japan, and one by E. W. Gorter for a synthetic substance in the laboratory.

The main points of Néel's four mechanisms will now be briefly discussed. The first and third only involve reversible physical changes while the second and fourth involve in addition irreversible physical and/or chemical changes. In his first mechanism he imagined a crystalline substance with two sub-lattices A and B with the magnetic moments of all the magnetic atoms in lattice B oppositely directed to those of lattice A. If the spontaneous magnetization of the two sets of atoms J_A and J_B varies differently with temperature, Néel suggested that the resultant magnetization of the whole, $J_A - J_B$, could reverse with change in temperature (see Fig. 7.5). E. W. Gorter and J. A. Schulkes two years later synthesized a range of substances with the properties predicted by Néel, although no naturally occurring rock has been found which behaves in this manner.

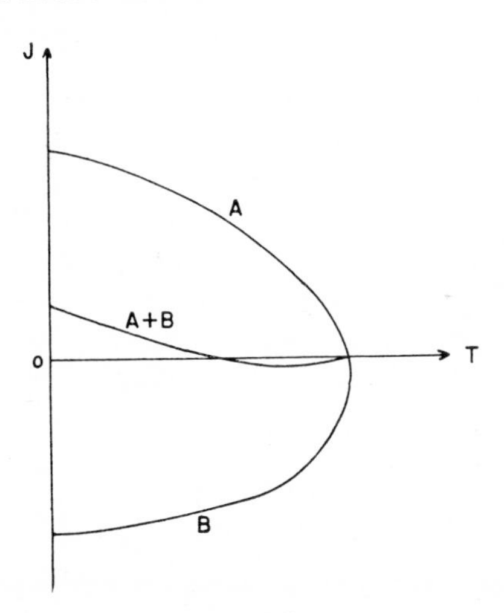

FIG. 7.5.

Néel's second mechanism is a modification of the first in which $J_A > J_B$ at all temperatures so that no reversal would take place. However, Néel suggested that subsequent to the formation of such a substance, chemical or physical changes might occur which would lead to the demagnetization of lattice A, leaving the reverse magnetization of lattice B predominant. No evidence of such a possibility actually occurring in nature has yet been found.

For his third mechanism, Néel considered a substance containing a mixture of two different types of grains A and B, one with a high Curie point T_A and a low intensity of magnetization J_A and the other with a low Curie point T_B and a high magnetization J_B (see Fig. 7.6). When such a substance cools from a high tempera-

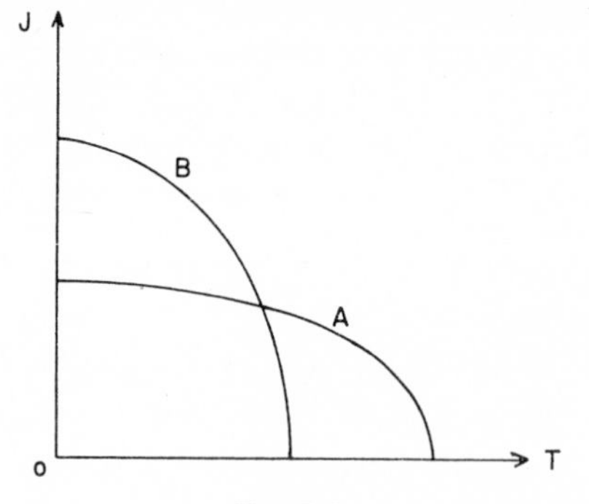

FIG. 7.6.

ture, substance A, because of its higher Curie point, magnetizes first in the direction of the ambient field. When the temperature falls below T_B, substance B becomes magnetic but will be subject to the dual influence of the ambient field and of the field due to the grains of substance A. Néel suggested that under suitable geometrical conditions, the resultant direction of magnetization of B can be on the average opposite to that of the ambient field. At room temperature the greater value of J_B makes the resultant magnetization of the whole in the opposite direction to the ambient field. In 1951 T. Nagata found a dacite pumice from Haruna in Japan which was reversely magnetized in the field and which behaved in the laboratory in the way which Néel had predicted. However, the great majority of igneous rocks which show reversal in the field do not show this reversal property in the laboratory.

Néel's fourth mechanism, like his second, involves the possibility of subsequent demagnetization by physical or chemical changes. Thus reverse magnetization might be possible later in time, even though initially the intensity of the B component was not large enough. Since discrete magnetized grains free to rotate must align themselves along and not against the field, only the second and fourth of Néel's mechanisms could apply to sedimentary rocks—igneous rocks could, in theory, become reversely magnetized by any of them.

Reversals occurred during the Pre-Cambrian and have been observed in all subsequent periods except the Permian. There is no evidence that periods of either polarity are systematically of longer or shorter duration. At least 15 successive groups of rocks with opposite polarity have been found in Iceland over a period of 30 million years and the total number may well exceed 30 taking the geological record as a whole. The length of normal and reversed magnetic periods during the Tertiary seems to be about one-half million years based on geologic estimates with an upper limit of two to three million years. The duration of the transition (estimated from geologic evidence) is several thousand years. A. Brynjolfsson (1957) has carried out some studies on the behaviour of the field during the short period ($\simeq$ 10,000 years) when the polarity is changing from one sign to another, but it is not possible to decide whether the reversal is due to a change in dipole moment, the axis remaining fixed, or through a movement of the dipole axis through 180°, the moment remaining unchanged.

The youngest reversely magnetized rocks are early Pleistocene or very late Pliocene. Although the exact ages of the rocks are subject to some uncertainty, the stratigraphic control is adequate to indicate that none of the late Pleistocene or recent rocks are reversely magnetized. The last appearance of reversely magnetized rocks at about the same time in such widely separated areas as Japan, Kenya, France, Iceland and Arizona strongly suggests field reversal.

An extremely interesting finding is that all rocks of Permian age have normal polarity (reversed magnetizations have been suggested for the Tartarskij sediments, but the scatter in the measurements is too great to be able to come to any definite conclusion). If the field reversal hypothesis is incorrect, it follows that mineral assemblages necessary for self-reversal are abundant in Carboniferous and Triassic rocks (both these periods have many reversals), but are missing in all Permian rocks. Such a conclusion is very difficult to believe—it is far more plausible to assume that the field did not alternate during the Permian.

E. Asami (1954) has examined some early Pleistocene lavas at Cape Kawajiri, Japan. Several hundred specimens were taken from closely spaced sites along the coastline. Along some stretches of the coast all the magnetization was normal; in other stretches it was reversed, and on some stretches normal and reversed were found close together. Such results show that one must be cautious about interpreting all reversals as due to a field reversal and the problem of deciding which reversed rocks indicate a reversal of the field may in some cases be extremely difficult. To prove that a reversed rock sample has been magnetized by a reversal of the Earth's field, it is necessary to show that it cannot have been reversed by any physico-chemical process. This is a virtually impossible task since physical changes may have occurred since the initial magnetization or may occur during certain laboratory tests. More definite results can only come from the correlation of data from rocks of varying type at different sites and by statistical analyses of the relation between the polarity and other chemical and physical properties of the rock sample. If the dipole field of the Earth has reversed it is most probably a result of physical processes occurring in the core of the Earth and should thus be quite uncorrelated with physical processes associated with the outer mantle or the atmosphere such as orogenic and volcanic activity or climatic changes.

If the origin of reversals is one of the instantaneous self-reversal mechanisms (such as that of the Haruna dacite), then normally and reversely magnetized rocks should be randomly distributed throughout a group of rocks of different ages. If reversals are due to one of the time-dependent self-reversal mechanisms, reversals should be increasingly abundant in older rocks. If, on the other hand, reversals are due to geomagnetic field reversals, normal and reversely magnetized groups of rocks should be exactly the same age over the entire Earth; and, unless it so happened that the Earth's field suffered more reversals in the past, the proportion of reversed magnetizations should not be greater among older rocks. Although there can be no doubt that self-reversal occurs in some rocks, the stratigraphic distribution of normally and reversely magnetized rocks strongly supports the field reversal hypothesis.

7.4 Continental Drift and the Growth of the Earth's Core

It is not appropriate in this book to discuss in detail the question of continental drift and convection in the Earth's mantle. However, these two subjects do have some bearing on the possible growth of the Earth's core and this aspect will be developed below.

A. Prey (1922) took values of the height of the land above and the depth of the oceans below sea-level, counting the latter negative, and expressed them as a series of spherical harmonics. He found the predominant term to be of degree $n = 1$, showing that the continents are concentrated in one hemisphere. The terms $n = 2$ and those greater than $n = 5$ are relatively weak, the terms $n = 3$, 4 and 5 being strong. Terms of odd degree have opposite signs at antipodal points, and Prey's analysis gives mathematical expression of the fact that continents are antipodal to oceans. F. A. Vening Meinesz (1952) argued that the positions of the continents today could result from a large scale, regular pattern of convective motions in the mantle, continental material tending to congregate at places where the currents are descending. Vening Meinesz and S. Chandrasekhar (1953) considered the Rayleigh problem of convection in a fluid contained in a spherical shell under a radial gravitational field. They showed that as the ratio of the radius of the inner to the outer boundary increases, the convective motion which is excited at marginal stability is characterized by harmonics of higher degree. Chandrasekhar showed that for a core of the present radius, the harmonics $n = 3$, 4 and 5 are almost equally likely to be excited at marginal stability. The positions of the continents before the most recent displacements occurred may be surmised from palaeomagnetic observations which indicate that at the end of the Palaeozoic Europe and North America were mainly in the northern hemisphere, in lower latitudes than at present. For such a continental distribution the $n = 4$ harmonic would be much more important relative to the $n = 5$ harmonic than at the present time. S. K. Runcorn (1962a) has thus concluded that before continental drift took place the positions of the continents were controlled essentially by a convective pattern of the $n = 4$ type and that this gave way during the past 200 million years to convection of the $n = 5$ type. The gradual growth of the core throughout the Earth's history as proposed by H. C. Urey (1952) could provide a simple explanation of why the convective pattern should change in this way. Runcorn (1962b) has estimated the growth of the Earth's core throughout geologic time, and found it to be linear for much of the life of the Earth, but that in comparatively recent times there has been a fundamental redistribution of convection currents in the mantle, due to a change in the dominant harmonics, resulting in continental drift.

The ratio η of the radius of the core to that of the Earth at which the Rayleigh number of a given harmonic becomes less than that of the harmonic of next lower degree and therefore may be expected to develop in place of the latter, may be calculated from Chandrasekhar's theory (1961). However, the results depend on the boundary conditions used. In the case of the Earth these are a free inner boundary and a rigid outer one, since shearing stresses must vanish at the core–mantle boundary and the crust of the Earth may be considered rigid. With these boundary conditions, S. K. Runcorn (1962b) showed that the ratio of the radius of the core to that of the Earth at which the second harmonic replaces the first as the mode at which instability occurs is 0·06; for the transition from the second to the third the value of η is 0·36, and for the third to the fourth it is 0·49. During these transitions the continents will be under considerable stress as the convection-pattern changes to a new form. These mode changes would be likely to cause world-wide orogeny, involving extensive recrystallization of crustal material. Radioactive age determinations of igneous and metamorphic rocks which have been formed in the deeper parts of the crust as the result of orogenic forces, have been found to be grouped in peaks around 200 m.y., 1000 m.y., 1800 m.y., and 2600 m.y. ago. The most recent peak covers much of the geological record since the middle Palaeozoic, and S. K. Runcorn (1962c) has argued that this is associated with continental drift and the transition from a fourth to a fifth degree convective pattern in the mantle. He also identifies the 1000-m.y. peak with the transition from the third to the fourth degree convective pattern, the 1800-m.y. peak with the

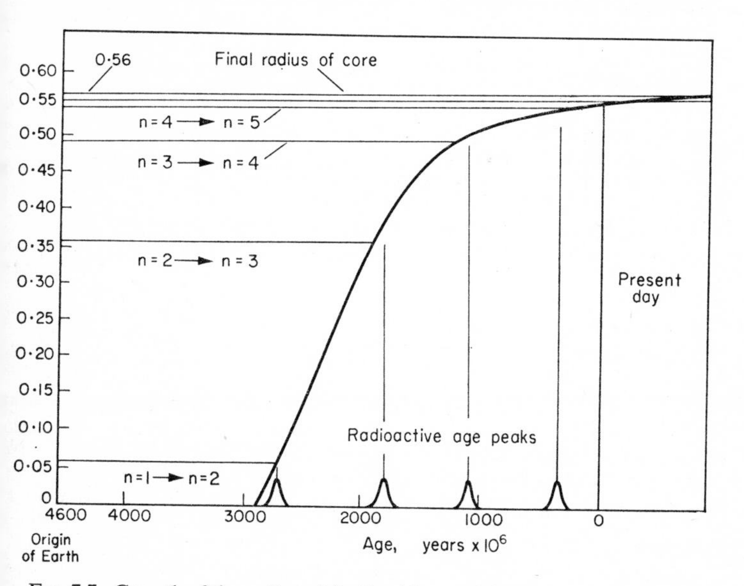

FIG. 7.7. Growth of the radius of the Earth's core (after S. K. Runcorn).

transition from the second to the third degree, and the 2600-m.y. peak with the transition from the first to the second. G. Gasti[l] has reviewed the radioactive age determinations and shown that the peaks are present whether the ages are determined by the rubidium–strontium method, the lead method, or the potassium–argon method, and that in spite of poor coverage in some continents, the peaks do appear to represent world-wide events. The dates at which the core attained these critical values are indicated in Fig. 7.7 (after S. K. Runcorn, 1962c) which shows that the core started its growth a little over 3000 m.y. ago. This is plausible since time must be allowed for the Earth to heat up to temperatures at which creep processes necessary for convection to occur become important.

References

No attempt will be made to give a complete coverage of all the recent wor[k] in the field of palaeomagnetism. The natural remanent magnetization of rock[s] is discussed in detail by S. K. RUNCORN (*Handbuch der Physik*, Band 4[7] 470–97, Springer-Verlag, 1956) with the object of determining the directio[n] of the geomagnetic field in the geological past. In addition two excellent revie[w] articles:

COX, A. and DOELL, R. R. Review of Palaeomagnetism. *Bull. Geol. So[c]. Am.* **71**, 645–768 (1960).

DOELL, R. R. and COX, A. Palaeomagnetism. *Advances in Geophysics* Vol. [8] pp. 221–313. Academic Press, New York (1961).

contain extensive bibliographies. An account of recent Soviet work is given b[y]

KALASHNIKOV, A. C. The history of the geomagnetic field. *Bull.* (*Izv.*) *Aca[d]. Sci. U.S.S.R. Geophys. Ser.*, 1243–79 (1961) (English ed. 1962, pp. 81[9]–38).

A list of Palaeomagnetic pole positions is published by E. Irving from tim[e] to time in the *Geophysical Journal*. The first list appears in **3**, 96–111 (1960).

ASAMI, E. On the reverse natural remanent magnetism of basalt of Ca[pe] Kawajiri, Yamaguchi Prefecture. *Jap. Acad. Proc.* **30**, 102–5 (1954).

ASAMI, E. Reverse and normal magnetization of the basaltic lavas at Kawajiri-misaki, Japan. *J. Geomag. Geoele.* **6**, 145–52 (1954).

BRUCKSHAW, J. M. and ROBERTSON, E. I. The magnetic properties of t[he] tholeiite dykes of North England. *Mon. Not. Roy. Astr. Soc. Geoph[ys]. Suppl.* 5, 308–20 (1949).

BRYNJOLFSSON, A. Studies of remanent magnetism and viscous magnetis[m] in the basalts of Iceland. *Adv. Phys.* **6**, 247–54 (1957).

BURLATSKAYA, S. P. The ancient magnetic field of the Earth. *Bull. (Izv.) Acad. Sci. U.S.S.R. Geophys. Ser.*, 524–8 (1962) (English ed. 1962, pp. 343–5).

CAMPBELL, C. D. and RUNCORN, S. K. Magnetization of the Columbia River basalts in Washington and Northern Oregon. *J. Geophys. Res.* **61**, 449–58 (1956).

CHANDRASEKHAR, S. The onset of convection by thermal instability in spherical shells. *Phil. Mag.* **44**, 233–41 (1953).

CHANDRASEKHAR, S. The onset of convection by thermal instability in spherical shells, A correction. *Phil. Mag.* **44**, 1129–30 (1953).

CHANDRASEKHAR, S. *Hydrodynamic and hydromagnetic stability*. Oxford (1961).

CHEVALLIER, R. L'aimantation des lavas de l'Etna et l'orientation du champ terrestre en Sicile du 12e au 17e Siècle. *Ann. Phys. Paris ser. 10*, **4**, 5–162 (1925).

CREER, K. M., IRVING, E. and RUNCORN, S. K. Palaeomagnetic investigations in Great Britain VI: Geophysical interpretation of palaeomagnetic directions from Great Britain. *Phil. Trans. Roy. Soc.* A., 250, 144–56 (1957).

GASTIL, G. The distribution of mineral dates in time and space. *Amer. J. Sci.* **258**, 1–35 (1960).

GRAHAM, J. W. The stability and significance of magnetism in sedimentary rocks. *J. Geophys. Res.* **54**, 131–67 (1949).

HOSPERS, J. Remanent magnetism of rocks and the history of the geomagnetic field. *Nature* **168**, 1111–12 (1951).

HOSPERS, J. Reversals of the main geomagnetic field. *Kon. Ned. Akad. Wet. Proc. of the Section of Sciences, Series B, Phys. Sci.* Part I: **56**, 467–76 (1953); Part II: **56**, 477–91 (1953); Part III: **57**, 112–21 (1954).

NÉEL, L. Some theoretical aspects of rock magnetism. *Adv. Phys.* **4**, 191–243 (1955).

PREY, A. Darstellung der Höhen und Tiefenverhaltnisse der Erde. *Abh. Ges. d. Wiss. Göttingen, Math. Phys. Kl. N.F.* **11**, 1 (1922).

ROCHE, A. Anomalies magnétiques accompagnant les massifs de pépérites de la Limagne d'Auvergne. *Compt. Rend. Acad. Sci.* **230**, 1603–4 (1950).

ROCHE, A. Sur les caractères magnétiques du système eruptif de Gergovie. *Compt. Rend. Acad. Sci.* **230**, 113–15 (1950).

ROCHE, A. Sur les inversions de l'aimantation rémanente des roches volcaniques dans les monts d'Auvergne. *Compt. Rend. Acad. Sci.* **230**, 1132–4 (1951).

ROCHE, A. Sur l'origine des inversions d'aimantation constantées dan les roches d'Auvergne. *Compt. Rend. Acad. Sci.* **236**, 107–9 (1953).

RUNCORN, S. K. Towards a theory of continental drift. *Nature* **193**, 311–14 (1962a).

RUNCORN, S. K. *Continental drift*, Chapter 1. Academic Press (1962b).

RUNCORN, S. K. Convection currents in the Earth's mantle. *Nature* **195**, 1248–9 (1962c).

THELLIER, É. and THELLIER, O. Sur l'intensité du champ magnétique terrestre dans le passe historique et géologique. *Ann. Géophys.* **15**, 285–376 (1959).

TORRESON, O. W., MURPHY, T. and GRAHAM, J. W. Magnetic polarization of sedimentary rocks and the Earth's magnetic history. *J. Geophys. Res.* **54**, 111–29 (1949).

UREY, H. C. *The Planets*. Yale Univ. Press (1952).

VENING MEINESZ, F. A. Convection currents in the Earth and the origin of the continents, I. *Kon. Ned. Akad. Weten.* **55**, 527–53 (1952).

Additional Reading

BLACKETT, P. M. S. *Lectures on Rock Magnetism*. Weizmann Science Press of Israel (1956).

NAGATA, T. *Rock Magnetism*. Maruzen, Tokyo (1953).

For an introductory account of the work on palaeotemperatures, see:

EMILIANI, C. Ancient temperatures. *Scientific American*, February (1958).

For an introductory account of the magnetic survey of part of the Pacific Ocean, see:

RAFF, A. D. The magnetism of the ocean floor. *Scientific American*, October (1961).

APPENDIX A

Magnetism

A.1 Introduction

Simple experiments with a bar magnet show that the magnetism is mainly concentrated in small regions near the ends of the bar, called the poles. If one end of a bar magnet is placed near the end of another which is mounted as a compass needle, it is found that the end of the compass needle is either attracted or repelled. Thus the two poles of a magnet have opposite polarity—the north (i.e. north seeking) pole is considered positive, and the south pole negative. If a bar magnet is broken into two parts, each part is found to be a complete magnet with positive and negative poles whose strength is unaltered by the breakage. Attempts have been made to separate positive and negative poles, but it has always been found that positive and negative poles appear on the two sides of the break. Thus, although positive and negative electric charges can exist separately, magnetic poles do not possess this independent existence. The basic entity in magnetism is a pair of poles, of opposite polarity (called a dipole).

The problem of obtaining "free" poles has been overcome, to a certain extent, by using long, thin, needle-shaped magnets, so that one pole is sufficiently far removed from the other as to be practically isolated from it. Experiment has then shown that the force due to a given pole varies inversely as the square of the distance from it. Thus the magnitude of the force between two poles of strength m_1 and m_2 situated a distance r apart is given by

$$\boldsymbol{F} = Cm_1m_2/r^2 \tag{A.1}$$

where the value of the constant of proportionality C depends upon the system of units and the magnetic property of the intervening medium. This equation is the basic equation of magnetostatics, and from it a unit pole can be defined. By analogy to electrostatics the concept of magnetic field, and magnetic potential can be defined. This approach to magnetism, using poles, although artificial, is extremely useful in many cases.

The magnetic force at a given point in a magnetic field is defined as the force which would be experienced by an isolated magnetic pole of unit strength placed at that point. This force can be expressed as a function of the work done in moving a unit pole against the field. Consider first the field due to a pole of strength m at the origin. The field is everywhere radial and at a distance r is a repulsion of magnitude

$$F = m/r^2 \tag{A.2}$$

If a unit pole is moved a distance dr against the field, external work dV must be supplied of amount

$$\mathrm{d}V = -F\mathrm{d}r$$

i.e.

$$\frac{\mathrm{d}V}{\mathrm{d}r} = -F = -m/r^2$$

Hence F can be expressed in terms of a scalar function V, called the magnetic potential. By integration

$$V = m/r + \text{constant} \tag{A.3}$$

The constant is usually taken as zero so that V is the work done in bringing unit pole from infinity to the point under consideration. The chosen value for the constant is, however, immaterial since it is only differences in the potential that are used.

If (X, Y, Z) are the components of the magnetic force $\mathbf{F}$ in the directions of the axes $0\,(x, y, z)$, it follows that

$$\left.\begin{aligned} X &= \frac{-\partial V}{\partial x} \\ Y &= \frac{-\partial V}{\partial y} \\ Z &= \frac{-\partial V}{\partial z} \end{aligned}\right\} \tag{A.4}$$

or

$$\mathbf{F} = -\nabla V \tag{A.5}$$

A.2 The Field of a Magnetic Dipole

Consider the potential due to a dipole composed of two poles of strength $-m$ at B and $+m$ at A where BA $= \delta s$ is the axis of the dipole and is in the direction **s**. Then the potential at a point P (r, θ) (see Fig. A.1) is given by

$$V = \frac{-m}{\text{PB}} + \frac{m}{\text{PA}}$$

$$\simeq \frac{-m}{r + \frac{1}{2}\delta s \cos\theta} + \frac{m}{r - \frac{1}{2}\delta s \cos\theta}$$

$$= m\delta s \cos\theta/(r^2 - \tfrac{1}{4}\delta s^2 \cos^2\theta)$$

$$\simeq \frac{m\delta s \cos\theta}{r^2}$$

since we may neglect the term in δs^2 in comparison with that in r^2

$$= \frac{M \cos\theta}{r^2} \tag{A.6}$$

where $M = m\,\delta s$ is the magnetic moment of the dipole. The radial and cross-radial components of force (in the directions r and θ increasing) are, from equation (A.5), given by

$$F_r = \frac{-\partial V}{\partial r} = \frac{2M\cos\theta}{r^3} \tag{A.7}$$

and

$$F_\theta = \frac{-1}{r}\frac{\partial V}{\partial \theta} = \frac{M\sin\theta}{r^3} \tag{A.8}$$

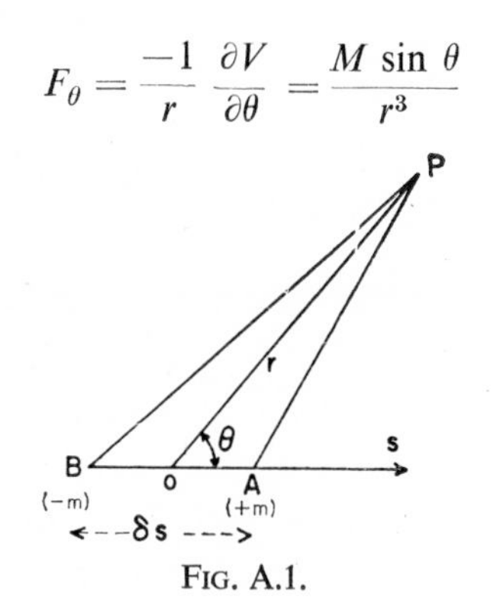

FIG. A.1.

A.3 Uniformly Magnetized Sphere

Any magnet can be supposed to be made up of a large number of small magnets or dipoles. Since a magnetic pole cannot be isolated, the magnetic moment of a dipole rather than its pole strength is the more fundamental concept. In any small element of volume $d\Omega$ of a magnetized body, the resultant moment of all the elementary dipoles may be effectively replaced by a single moment. We define the intensity of magnetization **J** as the moment per unit volume, i.e. $\mathbf{J}d\Omega$ is the resultant moment of all the elementary dipoles in the volume element $d\Omega$. In general, the vector **J** varies from point to point, but if it is constant in magnitude and direction, the body is said to be uniformly magnetized.

Consider a uniformly magnetized sphere of radius a, the intensity of magnetization **J** being in the direction **s**. Each element $d\Omega$ of the sphere may be regarded as a magnetic dipole of moment $\mathrm{J}d\Omega$. This in turn may be regarded as a pair of positive and negative poles $\pm \mathrm{J}d\Omega/\delta s$, a very small distance δs apart. The uniformly magnetized sphere is thus equivalent to a uniform distribution of positive poles $\mathrm{J}/\delta s$ per unit volume filling one sphere and an equal distribution of negative poles filling an equal sphere, displaced a very small distance δs from each other (see Fig. A.2). At points outside the sphere each distribution behaves

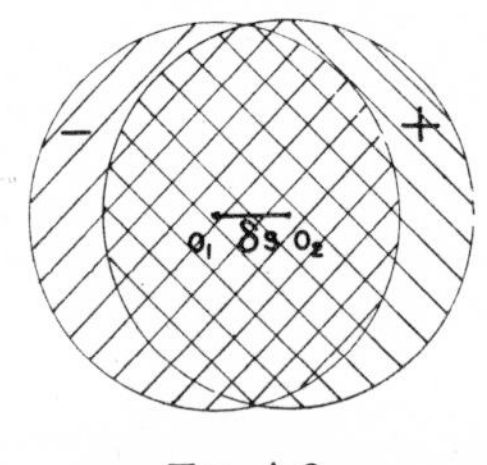

FIG. A.2.

as a concentrated pole of strength $\pm(\mathrm{J}/\delta s)(4/3)\pi a^3$, i.e. the field is that due to a dipole in which the pole strength is $(\mathrm{J}/\delta s)(4/3)\pi a^3$, the separation of the poles being δs. The moment of such a dipole is $M = (4/3)\pi a^3 \mathrm{J}$. The magnetic potential of a uniformly magnetized sphere at an external point is thus the same as that of a dipole situated at its centre, and is given by equation (A.6). This result could also have been obtained by solving the boundary value problem for the magnetic potential of a uniformly magnetized sphere, which outside the sphere satisfies Laplace's equation.

APPENDIX B

Spherical Harmonic Analysis

B.1 Introduction

Fourier showed in 1807 that any function $f(t)$ which is defined in the interval $t = 0$ to 2π and which satisfies certain conditions can be expressed as an infinite series of trigonometric functions.

i.e.
$$f(t) = a_0 + \sum_{n=1}^{\infty} (a_n \cos nt + b_n \sin nt) \tag{B.1}$$

In geophysics a more practical problem is that of approximating a function $f(t)$ by a finite series, i.e. of finding a nearly equal function $f_k(t)$ of the form

$$f_k(t) = a_0 + \sum_{n=1}^{k}(a_n \cos nt + b_n \sin nt) \tag{B.2}$$

If the coefficients are chosen so that the square of the difference $f(t) - f_k(t)$ averaged over the interval $(0, 2\pi)$ is a minimum, it can be shown that

$$\left.\begin{aligned} a_0 &= \frac{1}{2\pi}\int_0^{2\pi} f(t)\mathrm{d}t \\ a_n &= \frac{1}{\pi}\int_0^{2\pi} f(t) \cos nt \, \mathrm{d}t \\ b_n &= \frac{1}{\pi}\int_0^{2\pi} f(t) \sin nt \, \mathrm{d}t \end{aligned}\right\} \tag{B.3}$$

It can also be shown that for each value of n ($n \leqslant k$) the coefficients a_n and b_n are independent of k. By adding additional terms to the series $f_k(t)$, the approximation is in general improved but it does not alter the coefficients already obtained.

A set of functions $f_0(t)$, $f_1(t)$, $f_2(t)$ which are defined in an interval $t_1 < t < t_2$, form an orthogonal system if

$$\int_{t_1}^{t_2} f_n(t) f_m(t) \mathrm{d}t = 0 \qquad (n \neq m) \qquad \text{(B.4)}$$

Further the function $f_n(t)$ is said to be normalized if

$$\int_{t_1}^{t_1} f_n^2(t) \mathrm{d}t = 1$$

In deriving the expressions (B.3) use was made only of the orthogonality of the trigonometric functions. The Legendre functions $P_n(x)$ form an orthogonal set in the range $-1 < x < 1$ and may be used to approximate a given function in that range.

A spherical harmonic analysis may be regarded as a generalization of a Fourier analysis to three dimensions and gives an analytical representation of an arbitrary function of position on a sphere. By the substitutions (see Fig. B.1)

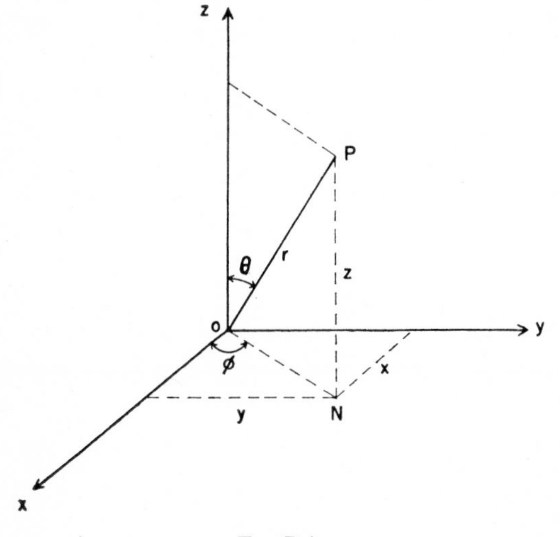

FIG. B.1.

$$\left.\begin{aligned} x &= r \sin\theta \cos\phi \\ y &= r \sin\theta \sin\phi \\ z &= r \cos\theta \end{aligned}\right\} \tag{B.5}$$

where $r > 0$, $0 \leqslant \theta \leqslant 180°$, $0 \leqslant \phi \leqslant 360°$, any function $f(x, y, z)$ can be expressed as a function $f(r, \theta, \phi)$. Thus any function of position on a sphere (radius a) is a function of θ and ϕ only, and we can write

$$f(\theta, \phi) = \sum_{m=0}^{\infty} \left\{ a_m(\theta) \cos m\phi + b_m(\theta) \sin m\phi \right\} \tag{B.6}$$

$$= \sum_{n=0}^{\infty} \sum_{m=0}^{n} \left\{ A_n{}^m \cos m\phi + B_n{}^m \sin m\phi \right\} P_n{}^m(\theta) \tag{B.7}$$

where $P_n{}^m(\theta)$ are the associated Legendre functions. The properties of these functions will not be discussed in this book. They may be found in any standard text on mathematical physics (e.g. H. and B. S. Jeffreys 1950, I. N. Sneddon 1956).

B.2 Spherical Harmonic Analysis of the Earth's Magnetic Field

Assuming that the Earth is a sphere of radius a and that there is no magnetic material near the ground, the Earth's magnetic field can be derived from a potential function V which satisfies Laplace's equation and can be represented as a series of spherical harmonics in the form

$$V = a \sum_{n=0}^{\infty} \sum_{m=0}^{n} P_n{}^m(\theta) \Big[\left\{ c_n{}^m (r/a)^n + (1 - c_n{}^m)(a/r)^{n+1} \right\} A_n{}^m \cos m\phi$$

$$+ \left\{ s_n{}^m (r/a)^n + (1 - s_n{}^m(a/r)^{n+1} \right\} B_n{}^m \sin m\phi \Big] \tag{B.8}$$

$c_n{}^m$ and $s_n{}^m$ are numbers lying between 0 and 1 and represent the fractions of the harmonic terms $P_n{}^m(\cos\theta) \cos m\phi$ and $P_n{}^m (\cos\theta) \sin m\phi$ in the expansion of V which on the surface of the sphere ($r = a$) are due to matter outside the sphere. It is also assumed that there are no electric currents flowing across the surface of the Earth—if there were they would set up a non-potential field and thus contribute a part of the Earth's magnetic field which could not be represented by equation (B.8). Such a field could be detected by the evaluation of line integrals of the horizontal magnetic force around closed contours on the Earth's surface, the value of such an integral being $4\pi J$ where J is the total electric current flowing through the contour. Such line integrals have been evaluated and do not vanish exactly—calculations by E. H. Vestine and J. Lange for the epoch 1945 indicate Earth–air currents of the order of 10^{-12}A/cm^2. However, their results have a completely random distribution and the non-zero value of the integral is most probably due to incomplete and imperfect magnetic data.

The potential V cannot of course be measured directly—what can be determined are the three components of force $X = (1/r)(\partial V/\partial \theta)$ (horizontal, northward), $Y = (-1/r \sin \theta)/(\partial V/\partial \phi)$ (horizontal, eastward) and $Z = \partial V/\partial r$ (vertical, downward) at the Earth's surface $r = a$.

Z (at $r = a$) may be expanded as a series of spherical harmonics

$$Z = (\partial V/\partial r) = \sum_{n=0}^{\infty} \sum_{m0}^{n} P_n^m(\cos \theta)\left\{ \alpha_n^m \cos m\phi + \beta_n^m \sin m\phi \right\} \quad \text{(B.9)}$$

and the coefficients α_n^m, β_n^m may be determined from the observed values of Z.

By differentiating equation (B.8) with respect to r and then writing $r = a$, we have

$$(\partial V/\partial r) = \sum_{n=0}^{\infty} \sum_{m=0}^{n} P_n^m(\cos \theta)\left[\left\{ nc_n^m - (n+1)(1 - c_n^m)\right\} A_n^m \times \right.$$

$$\left. \times \cos m\phi + \left\{ ns_n^m - (n+1)(1 - s_n^m)\right\} B_n^m \sin m\phi \right] \quad \text{(B.10)}$$

The coefficients of each separate harmonic term for each n and m must be equal in the two expansions of $\partial V/\partial r$ given in equations (B.9) and (B.10).

$$\left.\begin{aligned} \text{Hence} \qquad \alpha_n{}^m &= \left\{nc_n{}^m - (n+1)(1-c_n{}^m)\right\}A_n{}^m \\ \text{and} \qquad \beta_n{}^m &= \left\{ns_n{}^m - (n+1)(1-s_n{}^m)\right\}B_n{}^m \end{aligned}\right\} \tag{B.11}$$

Again from an analysis of the observed values of X and Y, the coefficients in the following two expansions derived from equation (B.8) may be obtained

$$Y_{r=a} = \left(\frac{-1}{r\sin\theta}\frac{\partial V}{\partial\phi}\right)_{r=a} = \frac{1}{\sin\theta}\sum_{n=0}^{\infty}\sum_{m=0}^{n} P_n{}^m(\cos\theta)\left[mA_n{}^m \sin m\phi - mB_n{}^m \cos m\phi\right] \tag{B.12}$$

$$X_{r=a} = \left(\frac{1}{r}\frac{\partial V}{\partial\theta}\right)_{r=a} = \sum_{n=0}^{\infty}\sum_{m=0}^{n} (\mathrm{d}/\mathrm{d}\theta)P_n{}^m(\cos\theta) \times \left[A_n{}^m \cos m\phi + B_n{}^m \sin m\phi\right] \tag{B.13}$$

Both these equations contain $A_n{}^m$ and $B_n{}^m$ and thus if values of X are known all over the world, values of Y can be deduced. If there is disagreement between observed and calculated values of Y, it would imply that the field was not completely derivable from a potential V and hence that Earth–air currents do exist. When Gauss first carried out such calculations in 1839 he found no discrepancy. From a knowledge of the coefficients $A_n{}^m$, $B_n{}^m$, $\alpha_n{}^m$ and $\beta_n{}^m$, equations (B.11) determine $c_n{}^m$ and $s_n{}^m$. Gauss showed from the data available at that time that $c_n{}^m = s_n{}^m = 0$, i.e. the cause of the Earth's magnetic field is entirely internal.

The coefficients of the field of internal origin are

$$g_n{}^m = (1 - c_n{}^m)A_n{}^m, \quad h_n{}^m = (1 - s_n{}^m)B_n{}^m \tag{B.14}$$

and are known as Gauss coefficients. If the external field is negligible, equations (B.14) reduce to $g_n{}^m = A_n{}^m$, and $h_n{}^m = B_n{}^m$. Values of these coefficients as obtained by different investigators since the time of Gauss are given in Table (B.1). It is clear that

Table B1 Spherical Harmonic Analysis of the Earth's Main Field.
(units 10^{-4} Γ)

Source	Epoch	g_1^0	g_1^1	h_1^1	g_2^0	g_2^1	h_2^1	g_2^2	h_2^2
Gauss	1835	−3235	−311	+625	+51	+292	+12	−2	+157
Erman–Petersen	1839	−3201	−284	+601	−8	+257	−4	−14	+146
Adams	1845	−3219	−278	+578	+9	+284	−10	+4	+135
Adams	1880	−3168	−243	+603	−49	+297	−75	+61	+149
Fritsche	1885	−3164	−241	+591	−35	+286	−75	+68	+142
Schmidt	1885	−3168	−222	+595	−50	+278	−71	+65	+149
Neumayer–Petersen	1885	−3157	−248	+603	−53	+288	−75	+65	+146
Dyson–Furner	1922	−3095	−226	+592	−89	+299	−124	+144	+84
Afanasieva	1945	−3032	−229	+590	−125	+288	−146	+150	+48
Vestine–Lange	1945	−3057	−211	+581	−127	+296	−166	+164	+54
Finch–Leaton	1955	−3055	−227	+590	−152	+303	−190	+158	+24

by far the most important contribution to V comes from the term containing g_1^0, which is proportional to P_1 (cos θ)/r^2, i.e. cos θ/r^2 and corresponds to the field of a uniformly magnetized sphere.

An outstanding feature of Table (B.1) is the secular variation of the individual coefficients which is clearly apparent in spite of individual scatter. There appears to have been an overall decrease in the dipole moment of the Earth's field of about 6 per cent during the past century. Over the same time interval, the inclination of the dipole axis appears to have remained sensibly constant at about 11·6°, although a slow and somewhat irregular precession of the axis appears to have taken place from east to west. The non-dipole components of the field are subject to strong and comparatively rapid secular variations with apparently no constant components.

Suggestions for Further Readings

CHAPMAN, S. and BARTELS, J. *Geomagnetism.* Oxford Univ. Press (1940). (In particular Vol. II, Chapters XVI, XVII and XVIII.)

JEFFREYS, H. and JEFFREYS, B. S. *Methods of Mathematical Physics.* Cambridge Univ. Press (1950).

SNEDDON, I. N. *Special Functions of Mathematical Physics and Chemistry.* Oliver & Boyd (1956).

APPENDIX C

Equations of the Lines of Force of a Uniformly Magnetized Sphere

THE lines of force by symmetry lie on surfaces of revolution about the axis and in any axial plane are given by

$$\frac{dr}{Z} = \frac{r d\theta}{H} \tag{C.1}$$

i.e. from equations (A.7) and (A.8)

$$\frac{dr}{2M \cos\theta/r^2} = \frac{r d\theta}{M \sin\theta/r^2} \tag{C.2}$$

$$\frac{dr}{2\cos\theta} = \frac{r d\theta}{\sin\theta}$$

or (C.3)

$$\frac{dr}{r} = \frac{2\cos\theta}{\sin\theta} d\theta$$

which integrates to give

$$\ln r = 2 \ln \sin\theta + \text{constant}$$

or (C.4)

$$r = C \sin^2\theta$$

where different values of C correspond to different lines of force.

Equation (C.4) may be written in the form

$$r/a = \sin^2\theta/\sin^2\theta_0 \tag{C.5}$$

where θ_0 is the value of θ at which the line of force meets the sphere $r = a$. The maximum distance of a line of force from the sphere is obtained when $\theta = 90°$, i.e. above the equator, when $r = r_{max} = a/\sin^2\theta_0$. Thus if $\theta_0 = 30°$, $r_{max} = 4a$ corresponding to a height $3a$ above the equator.

APPENDIX D

Magnetohydrodynamics

D.1 Dimensional Analyses

The difficulties in those problems of magnetohydrodynamics which are of geophysical interest are mainly mathematical, since the equations are non-linear. A particularly useful approach is to consider the orders of magnitude of the different terms and hence determine those factors which are likely to be the most important. Some insight into the physical situation may then be obtained by solving an idealized linearized problem. This can be illustrated by the problem of thermal convection in a layer of fluid heated from below and subject to the combined effects of rotation and a magnetic field.

Consider in the first place the case when there is neither rotation nor a magnetic field. On account of thermal expansion the fluid above, being colder, will be heavier. This is an unstable situation, and if the temperature gradient is sufficiently adverse, instability sets in as a pattern of cellular convection. Rayleigh (1916) showed that the criterion is the numerical value of the non-dimensional number

$$R_a = \frac{g\alpha\beta d^4}{\kappa\nu} \tag{D.1}$$

where g is the acceleration due to gravity, d the thickness of the layer, β the temperature gradient, and α, κ, ν the coefficients of volume expansion, thermal diffusivity and kinematic viscosity respectively. R_a is known as the Rayleigh number and only when it exceeds about 1000 (the exact number depending on the nature of the boundaries) does the fluid convect. It is clear that higher temperature gradients may be withstood before instability sets in in a fluid of higher viscosity and/or higher thermal conductivity.

When the fluid rotates with angular velocity $\mathbf{\Omega}$ about the vertical, it is subject, in addition to gravity, to the Coriolis force. It is found that this inhibits the onset of convection—the extent depending on the value of the non-dimensional number (the Taylor number)

$$T_a = \frac{4\Omega^2 d^4}{\nu^2} \tag{D.2}$$

The value of T_a at which rotation begins to make itself felt is about 1000. Vortex lines have a tendency to be dragged along with the fluid, the attachment of the fluid to the vortex lines being stronger the lower the viscosity and the higher the angular velocity. Thus as Ω increases and/or ν decreases, motions at right angles to the vertical become increasingly difficult, preventing an easy closing in of the stream lines required for convection.

The effect of a magnetic field is also to inhibit convection—the effect being greater the stronger the magnetic field $\mathbf{H}$ and the higher the electrical conductivity σ. When the field is strong (or the conductivity high) the lines of force tend to be glued to the material making motions at right angles to the vertical increasingly difficult. The critical Rayleigh number for the onset of instability in the presence of a magnetic field depends on the value of the non-dimensional number.

$$Q = \frac{\sigma\mu^2 H^2 d^2}{\rho\nu} \tag{D.3}$$

where μ and ρ are the permeability and density respectively. When T_a is zero, it is found that the critical Rayleigh number increases indefinitely with Q, although no appreciable effect is felt until Q exceeds about 1000.

When the motion is subjected to viscous, Coriolis and magneto-hydrodynamic forces, the situation is very complex. If viscous forces may be neglected ($T_a > 10^3$, $Q > 10^3$) it is found that rotation is more or less influential than the magnetic field according as

$$\frac{(T_a)^{\frac{1}{2}}}{Q} = \frac{2\Omega\rho}{\sigma\mu^2 H^2} \gtrless 1 \tag{D.4}$$

It is instructive to estimate the order of magnitude of the terms in the hydrodynamical equation of fluid motion in the Earth's core. The Navier–Stokes equation is

$$\rho\left\{\frac{\partial \mathbf{v}}{\partial t} + (\mathbf{v}\,.\,\nabla)\mathbf{v} + 2\mathbf{\Omega}\times\mathbf{v} - \nu\nabla^2\mathbf{v}\right\} - \frac{\mu}{4\pi}\,\mathrm{curl}\,\mathbf{H}\times\mathbf{H}$$
$$= -\,\nabla p + \rho\nabla \mathrm{W} \tag{D.5}$$

where $\mathbf{v}$ is the velocity relative to a system rotating with an angular velocity $\mathbf{\Omega}$, $\mathbf{H}$ is the magnetic field, p the pressure, W the gravitational potential (in which is absorbed the centrifugal force) and ρ, ν, μ are the density, kinematic viscosity and permeability respectively. Let V, L and H be typical values of a velocity, distance scale and magnetic field strength. Replacing ∇ by L^{-1} and $\partial/\partial t$ by V/L, we may estimate the magnitudes of the different terms on the left-hand side of equation (D.5) as shown in Table D.1.

It does not follow automatically that because one term in Table D.1 is numerically greater than another that it is dynamically more important. (Turbulence sets in in a straight pipe when the Reynolds Number R_e = inertial force/viscous force = VL/ν exceeds about 1000.) It seems safe, however, to neglect the inertial term compared to the Coriolis term. Their ratio = $V/2L\Omega$ is called the Rossby number and in the core has a value of $\simeq 3 \times 10^{-6}$. In the atmosphere, the Rossby number is as high as 10^{-1}, and the Coriolis force certainly dominates inertial forces. It also seems safe to neglect the viscous term compared to the Coriolis term, since the Taylor number T_a = (Coriolis force/viscous force)^2 well exceeds 1000. (From Table D.1, $10^8 \ll T_a < 10^{32}$ in the Earth's core.) It is not possible, however, to assess the relative influence of the Coriolis and electromagnetic forces. Chandrasekhar's criterion (see equation D.4) yields a value of the ratio $(T_a)^{\frac{1}{2}}/Q$ of the order of unity.

TABLE D.1

(*after R. Hide, 1956*)

	Inertial force	Coriolis force	Viscous force	Electromagnetic pondero-motive force
Term in equation	$\rho\frac{\partial \mathbf{v}}{\partial t}$, $\rho(\mathbf{v} \cdot \nabla)\mathbf{v}$	$\rho 2\Omega \times \mathbf{v}$	$\rho\nu\nabla^2\mathbf{v}$	$\frac{\mu}{4\pi}$ curl $\mathbf{H} \times \mathbf{H}$
Order of magnitude	$\rho V^2/L$, $\rho V^2/L$	$2\rho\Omega V$	$\rho\nu V/L^2$	$\mu H^2/4\pi L$
Numerical value	3×10^{-10}, 3×10^{-10}	10^{-4}	$10^{-8} \sim 10^{-20}$	7×10^{-7}

L is taken as 3×10^8 cm, the size of a typical region of secular variation activity.

H is taken as 50 Γ (the strength of Bullard's toroidal field).

$\Omega = 7 \times 10^{-5}$ rad/sec.

V is taken as 0·1 cm/sec (the average value of the westward drift, and the value of the zonal component of flow in Bullard's dynamo theory).

$\rho \simeq 10$ g/cm^3.

ν is difficult to estimate with any certainty. Values proposed lie in the range $10^{-3} < \nu < 10^9$ cm^2/sec. The latter is certainly an extreme upper limit and it seems probable that $\nu < 10^4$ cm^2/sec. Most conclusions are insensitive to the value chosen, however.

$\mu \simeq 1$ (e.m.u.), since the core is not likely to be ferromagnetic at the temperatures which prevail there.

D.2 The Basic Equations of Magnetohydrodynamics in the Earth's Core

W. M. Elsasser (1954) has shown by a detailed dimensional analysis that in all geophysical and astrophysical problems the displacement current and all purely electrostatic effects are negligible, as are all relativistic effects of order higher than v/c where $\mathbf{v}$ is the fluid velocity. Thus the electromagnetic field equations are the usual Maxwell equations

$$\text{curl}\,\mathbf{E} = -\mu\frac{\partial\mathbf{H}}{\partial t} \tag{D.6}$$

$$\text{curl}\,\mathbf{H} = 4\pi\mathbf{J} \tag{D.7}$$

$$\text{div}\,\mathbf{H} = 0 \tag{D.8}$$

where $\mathbf{H}$ and $\mathbf{E}$ are the magnetic and electric fields respectively, and $\mathbf{J}$ the electric current density. The magnetic permeability μ and electrical conductivity σ will be assumed to be constant. The electromotive forces which give rise to $\mathbf{J}$ are due both to electric charges and to motional induction so that the total current $\mathbf{J}$ is given by

$$\mathbf{J} = \sigma(\mathbf{E} + \mu\mathbf{v} \times \mathbf{H}) \tag{D.9}$$

Taking the curl of equation (D.7) and using equations (D.9) and (D.6), $\mathbf{E}$ can be eliminated, leading to the equation

$$\text{curl curl}\,\mathbf{H} = 4\pi\mu\sigma(-\partial\mathbf{H}/\partial t + \text{curl}\,\mathbf{v} \times \mathbf{H}) \tag{D.10}$$

Since $\text{curl curl}\,\mathbf{H} = \text{grad div}\,\mathbf{H} - \nabla^2\mathbf{H} = -\nabla^2\mathbf{H}$ on using (D.8), we finally obtain

$$(\partial\mathbf{H}/\partial t) = \text{curl}\,(\mathbf{v} \times \mathbf{H}) + \nu_m\nabla^2\mathbf{H} \tag{D.11}$$

where

$$\nu_m = 1/4\pi\sigma\mu \tag{D.12}$$

is the "magnetic diffusivity" (in e.m.u.). Taking $\sigma = 3 \times 10^{-6}$, in the Earth's core, $\nu_m \simeq 3 \times 10^4\ \text{cm}^2/\text{sec}$.

Equations (D.8) and (D.11) give the relations between $\mathbf{H}$ and $\mathbf{v}$ which have to be satisfied from electromagnetic considerations. To these must be added the equation of hydrodynamics, viz. the Navier–Stokes equation (D.5) together with the equation of continuity, which for an incompressible fluid (the speed of flow

$\mathbf{v}$ is much less than the speed of sound in the Earth's core) reduces to

$$\text{div } \mathbf{v} = 0 \qquad \text{(D.13)}$$

Equations (D.11) and (D.5) contain only the vectors $\mathbf{v}$ and $\mathbf{H}$ and are the basic equations of field motion. These equations contain non-linear terms of three kinds, the term curl $(\mathbf{v} \times \mathbf{H})$ representing electromagnetic induction, the electromagnetic term $\mu/4\pi$ curl $\mathbf{H} \times \mathbf{H}$, and the inertia term $\rho(\mathbf{v} \cdot \nabla)\mathbf{v}$. As already mentioned it is this non-linearity which makes a complete mathematical solution next to impossible. No amount of effort will exhibit all the features of the hydromagnetic equations in a linearized approximation. It is possible, however, as has already been shown, to obtain some qualitative results from dimensional considerations.

If the material is at rest, equation (D.11) reduces to

$$(\partial \mathbf{H}/\partial t) = \nu_m \nabla^2 \mathbf{H} \qquad \text{(D.14)}$$

This has the form of a diffusion equation, and indicates that the field leaks through the material from point to point. Dimensional arguments indicate a decay time of order $L^2/\nu_m = 4\pi\mu\sigma L^2$ where L is a length representative of the dimensions of the region in which current flows. For conductors in the laboratory this is very small, rarely exceeding a few milliseconds—even for a copper sphere of radius 1 m it is less than 10 sec. For cosmic conductors, on the other hand, because of their enormous size, it can be very large. W. M. Elsasser has estimated the time of free decay of the Earth's field to be of the order of 15,000 years, supposing the core to consist of molten iron. T. G. Cowling has estimated that for the magnetic field of a sunspot the decay time is at least 300 years, and for a general solar magnetic field it is 10^{10} years. For a field in interstellar space in the galaxies it is still far greater.* It is clear that lines of magnetic force in large conducting masses leak only very slowly through the material.

---

* These extremely long decay times may be substantially reduced in stellar hydromagnetism since transport rates and free-decay coefficients may be increased by a large factor due to eddy diffusion.

As an alternative limiting case, suppose that the material is in motion but has negligible electrical resistance. Equation (D.11) then becomes

$$(\partial \mathbf{H}/\partial t) = \text{curl}\,(\mathbf{v} \times \mathbf{H}) \tag{D.15}$$

This equation is identical to that satisfied by the vorticity in the hydrodynamic theory of the flow of non-viscous fluids where it is shown that vortex lines move with the fluid. Thus equation (D.15) implies that the field changes are the same as if the magnetic lines of force are constrained to move with the material, i.e. the lines of force are "frozen" into the material.

When neither term on the right-hand side of equation (D.11) is negligible, both the above effects are observed, i.e. the lines of force tend to be carried about with the moving fluid and at the same time leak through it.

If L, T, V represent the order of magnitude of a length, time and velocity respectively, transport dominates leak if $LV \gg \nu_m$. The condition for the onset of turbulence in a fluid is that the non-dimensional Reynolds number $R_e = LV/\nu$ be numerically large. By analogy, a magnetic Reynolds number R_m may be defined as

$$R_m = LV/\nu_m \tag{D.16}$$

Thus the condition for transport to dominate leak is that $R_m \gg 1$. This condition is only rarely satisfied in the laboratory—in cosmic masses, however, it is easily satisfied because of the enormous size of L. Thus under laboratory conditions, lines of force slip readily through the material—in cosmic masses, on the other hand, the leak is very slow and the lines of force can be regarded as very nearly frozen into the material.

The viscous term $\rho\nu\nabla^2\mathbf{v}$ in equation (D.5) is of order $1/R_e$ compared to the inertia term $\rho(\mathbf{v} \cdot \nabla)\mathbf{v}$ and the dissipative term $\nu_m\nabla^2\mathbf{H}$ in equation (D.11) is of order $1/R_m$ compared to the other terms, further justifying the term magnetic diffusivity and magnetic Reynolds number.

To appreciate further the physical significance of R_m consider an electric conductor in which a system of currents is flowing.

In the absence of electromagnetic forces such currents will decay exponentially. From equation (D.11), on writing $\mathbf{v} = 0$, it follows that the decay time is of order $\tau = L^2/\nu_m$, so that from equation (D.16)

$$R_m = \frac{\tau V}{L} = \frac{\tau}{T}$$

where T refers to the mechanical motions of the fluid. The fact that R_m is numerically large corresponds to very large spontaneous decay times of electric currents and magnetic fields as compared to mechanical periods. Thus an essential feature of cosmic hydromagnetism is the fact that fluid motions may be considerable during a time in which the decay of the electromagnetic fields is quite small.

References

CHANDRASEKHAR, S. *Hydrodynamic and hydromagnetic stability*. Oxford (1961).

COWLING, T. G. *Magnetohydrodynamics*. Interscience (1957).

ELSASSER, W. M. Dimensional values in magnetohydrodynamics. *Phys. Rev.* **95**, 1–5 (1954).

ELSASSER, W. M. Hydromagnetism, I: A Review. *Amer. J. Phys.* **23**, 590–609 (1955).

HIDE, R. The hydrodynamics of the Earth's core. *Physics and Chemistry of the Earth*, pp. 94–137. Pergamon Press (1956).

APPENDIX E

Thc Vector-Wave Equation

In the absence of motion the magnetic field satisfies the differential equation (D.11) with $\mathbf{v} = 0$
i.e.

$$\nu_m \nabla^2 \mathbf{H} = (\partial \mathbf{H}/\partial t) \tag{E.1}$$

with suitable boundary conditions. Normal modes may be obtained by postulating that the field decays without changing shape,

$$\mathbf{H}(\mathbf{r}, t) = \mathbf{H}(\mathbf{r}) \exp(-\lambda t) \tag{E.2}$$

Defining k by the relation

$$\lambda = k^2 \nu_m \tag{E.3}$$

and assuming that both λ and k are real, equation (E.1) becomes the vector wave equation

$$\nabla^2 \mathbf{H} + k^2 \mathbf{H} = 0 \tag{E.4}$$

It is not possible to extend the familiar boundary-value theory of the scalar wave equation

$$\nabla^2 \psi + k^2 \psi = 0 \tag{E.5}$$

to its vector analogue (E.4).

A set of orthonormal modes may be constructed by imposing on the solutions of (E.5) linear boundary conditions for a boundary of arbitrary shape. A similar general theory for equation (E.4) has not been found, although the case of cylindrical and spherical vector waves has been worked out—see, for example, J. A. Stratton (1941). We will consider spherical waves. The components of $\nabla^2 \mathbf{H}$ in curvilinear coordinates differ from the expression $\nabla^2 H_r$ etc. and are obtained from the vector identity

$$\text{curl curl } \mathbf{H} = \text{grad div } \mathbf{H} - \nabla^2 \mathbf{H} \tag{E.6}$$

Any vector field defined in all space can be uniquely decomposed into a longitudinal (i.e. irrotational) and a transverse (i.e. divergence free) part. For a finite, bounded volume this analysis becomes more difficult, but a generalization of the procedure can be established (see, for example, E. N. Parker 1955).

It is easy to construct longitudinal (curl $\mathbf{H} = 0$) solutions of (E.4) given a solution of the scalar equation (E.5). Such a solution is

$$\mu = \nabla\psi \tag{E.7}$$

To obtain transverse solutions (div $\mathbf{H} = 0$) which are of most interest to us, it is clear that with any solution of (E.4), its curl is also a solution. Transverse solutions can be constructed in pairs which are each other's curl. Thus if div $\mathbf{S}$ = div $\mathbf{T} = 0$, then writing

$$k\mathbf{T} = \text{curl}\,\mathbf{S},\; k\mathbf{S} = \text{curl}\,\mathbf{T} \tag{E.8}$$

it follows by elimination that both $\mathbf{S}$ and $\mathbf{T}$ satisfy the vector wave equation (E.4).

Let ψ be a solution of the scalar wave equation (E.5), and write

$$\mathbf{T} = \text{curl}\,(\psi\mathbf{r}) = \nabla\psi \times \mathbf{r} \tag{E.9}$$

From equation (E.8) it is easily found that

$$\mathbf{S} = k\psi\mathbf{r} + k^{-1}\nabla[(\partial/\partial r)r\psi] \tag{E.10}$$

From equation (E.9), the components of the vector $\mathbf{T}$ are

$$(\mathbf{T})_r = 0, \qquad (\mathbf{T})_\theta = \frac{1}{\sin\theta}\frac{\partial\psi}{\partial\phi}, \quad (\mathbf{T})_\phi = -\frac{\partial\psi}{\partial\theta} \tag{E.11}$$

This type of transverse mode is called toroidal, and lies entirely in surfaces of constant r. Again from (E.10)

$$\left.\begin{aligned}(\mathbf{S})_r &= kr\psi + k^{-1}\frac{\partial^2}{\partial r^2}(r\psi)\\ (\mathbf{S})_\theta &= (kr)^{-1}\frac{\partial^2}{\partial r\partial\theta}(r\psi)\\ (\mathbf{S})_\phi &= (kr\sin\theta)^{-1}\frac{\partial^2}{\partial r\partial\phi}(r\psi)\end{aligned}\right\} \tag{E.12}$$

This type of transverse mode is called poloidal. If we have

rotational symmetry, $(\mathbf{T})_r = (\mathbf{T})_\theta = 0$ and $(\mathbf{S})_\phi = 0$. If these vectors represent magnetic fields, the field lines for a toroidal field are circles about the axis, while for a poloidal field, the field lines lie in meridional planes. For the scalar generating function ψ, we can take

$$\left.\begin{aligned} \psi_n{}^{ms} &= R(r)P_n{}^m(\cos\theta)\sin m\phi \\ \psi_n{}^{mc} &= R(r)P_n{}^m(\cos\theta)\cos m\phi \end{aligned}\right\} \quad \text{(E.13)}$$

where $R(r)$ is a function of r only and $P_n{}^m(\cos\theta)$ the associated Legendre functions.

That **S**, **T** are the only transverse solutions of the vector wave equation does not seem to have been proved, nor has it been emphasized that the ψ associated with **S** and the ψ associated with **T** are in general different. Also, poloidal and toroidal fields are often used, as here, in problems where the vector wave equation no longer describes what is happening. These difficulties may be overcome by the use of a theorem due to G. Backus (1958). It states that if a vector field **H** is transverse in a region, then for every choice of origin there exist unique scalars S and T such that

$$\mathbf{H} = -\operatorname{curl}\operatorname{curl}\mathbf{r}S - \operatorname{curl}\mathbf{r}T \quad \text{(E.14)}$$

where S and T average to zero on every spherical surface concentric with the origin.

In the case of magnetic diffusion the equation to be satisfied is

$$\nu_m \,.\, \operatorname{curl}\operatorname{curl}\mathbf{H} = -(\partial\mathbf{H}/\partial t)$$

which reduces, as Backus shows, to the two equations

$$\nu_m \,.\, \nabla^2 S = (\partial S/\partial t), \qquad \nu_m \nabla^2 T = (\partial T/\partial t)$$

The well-known classical methods may be used to solve these.

Often the equation or equations being considered are more complicated than the magnetic diffusion equation or have boundary conditions which are not easily reducible to conditions on S and T. The procedure is then to expand S and T in series of orthogonal functions in two of the three coordinates.